UNIVALENT FUNCTIONS

Volume I

A. W. Goodman
UNIVERSITY OF SOUTH FLORIDA

UNIVALENT FUNCTIONS

Volume I

MARINER PUBLISHING COMPANY, INC.

Printed in the United States of America

ISBN: 0-936166-10-X (Volume I)
ISBN: 0-936166-12-6 (2 Volume Set)

To the memory of

Ze'ev Jabotinsky

a true hero of Israel

PREFACE

1. THE PURPOSE AND PLACE OF THIS BOOK

This book is not a text, nor a monograph, nor an encyclopedia, but it is a combination of these three types. It begins as an elementary textbook, and indeed the first eight chapters together with Chapters 10 and 11 should constitute a very nice semester course on univalent functions. All of the chapters conclude with a set of problems, and wherever it is appropriate we give answers or solutions in the back of the book. Some of these problems may seem trivial to the expert, and indeed they are; but this book is aimed initially at the student, and what may be trivial to the expert is exactly what the beginner needs to progress toward becoming an expert.

Every book bridges a gap—a gap between the more elementary books and the deeper ones on the same subject. However, it seems to the author that in the theory of univalent functions, the gap that needs bridging is far wider than in other branches of mathematics. On one side of the gorge we have the various textbooks which either ignore the theory of univalent functions or devote only a few pages to the topic. The most extensive treatment of the subject by a textbook may be found (A) in the three-volume text of Markushevich [1] and (B) in the two-volume work of Sansone and Gerretsen 1969 [2-B-1263][1].

On the other side of the gorge are the six advanced books: (A) *Univalent Functions and Conformal Mapping* by J. Jenkins 1958 [1-B-618], (B) *Univalent Functions–Selected Topics* by Glenn Schober 1975 [2-B-1280], (C) *Univalent Functions* by C. Pommerenke 1975 [2-B-1168], (D) *Coefficient Regions for Schlicht Functions* by A. C. Schaeffer and D. C. Spencer 1950 [1-B-1240], (E) *Multivalent Functions* by W. K. Hayman 1958 [1-B-515], and (F) *Geometric Function Theory* by G. M. Goluzin 1966 [2-B-464] that is available in Russian, German, and English. At this time several books on the same topic are in preparation, but it seems likely that these books will also treat the elementary portions of the theory quite quickly in order to have more space for the deeper and more difficult material.

Our first objective is to cover the elementary parts of the theory of univalent functions in a leisurely and detailed fashion. In this way we hope to attract the beginning graduate student by explaining carefully those items which other textbooks slide over all too quickly. At the same time, this book may have some appeal for the mathematician who has already specialized in some other field but wishes to learn something about univalent functions in a painless way.

Our second objective is to attempt (foolishly) a reasonably complete report on the work that has been done in the field so that a Ph. D. student looking for a thesis problem may obtain some guidance and help. At this point we should mention the *Bibliography of Schlicht (Univalent) Functions*,[2] compiled by S. D. Bernardi, that lists all the publications in this specialty that appeared before 1981. The first part lists 1,694 papers and books (prior to 1966), the second part gives another 1,563 such items (1966 to 1976), and the third part includes 1,025 items (1976 to 1981). Thus it is obvious that even a ten-volume "Introduction" could not adequately cover the field.

To handle this torrent of papers, we often shift from a textbook style to that of an encyclopedia, merely stating the theorems and citing the proper sources for the student who wishes to see the proof. This shift

[1] For an explanation of this notation, see the next section.

[2] Bernardi's bibliography is now available from Mariner Publishing Co., Inc.

in form reaches a high point in Chapter 13, a chapter that does not contain a sinlge proof.

Finally, we try on occasion to point out open problems that the reader may wish to attack. Here we are on very dangerous ground. First, among the more than 4,000 papers on the subject (of which the author has only about 2,000 reprints) there may be one (missed by the author) which already contains the solution to a problem listed in this book as unsolved. Second, papers on univalent functions continue to appear all over the world, and a problem that was open when this book was written may well be solved by the time the student reads it here.

2. SOME NOTES ON FORM

A well-written book contains at the end a list of references. In our case such a list might consume another fifty pages. Since S. D. Bernardi has devoted so much time and effort to compiling an exhaustive bibliography, we will economize on space by using it. Thus, a theorem or definition followed by "J. Doeskovič 1936 [1-B-1313]" means that Doeskovič gave the theorem or definition in 1936 and the exact reference (title, journal, etc.) can be found in Part I of Bernardi's bibliography as item 1313. The notation [1-B-A-44] means that the reference is contained in the Appendix of Part I. Similarly, [2-B-1001] means that the exact reference is item 1001 in Part II of Bernardi's bibliography. Papers or books not listed in Bernardi's bibliography are handled in the usual way, [1], with the exact reference at the end of the book.

It is clear that too many symbols can spoil the book, and we make some small effort at preventing them from taking command. But a symbol such as S which replaces "the set of functions $f(z)$ regular and univalent in the unit disk for which $f(0) = 0$ and $f'(0) = 1$" certainly saves us a tremendous amount of space. To help the reader we include in the back of the book a table of the special symbols used in this book.

It is standard practice to use a single letter to denote a set of function. However, the number of interesting classes has proliferated so that it is now near to or exceeds 26. Consequently we take the liberty of using double letters. Further this may avoid some of the confusion now current, since some authors use K for convex functions and C for close-to-convex functions, while others interchange the meanings of these two letters. Here we use the letters CV and CC for these two sets of functions, respectively. Once the barrier of one letter has been broken, we can use any number of letters to denote a set whenever it seems desirable to do so.

We can save space and make the book more readable if we omit some of the more obvious assumptions. For example, every function considered in this book is regular (analytic) in D, the domain under consideration (usually E), or has at most a finite number of poles. Rather than state such an assumption in every theorem and definition, we prefer to omit it whenever the context makes clear what we are assuming about $f(z)$. On occasion we do insert the phrase "let $f(z)$ be regular and univalent in E," but more in the spirit of a gentle reminder than as a dogmatic duty. If the function has a pole, we will of course state this fact.

In numbering equations, theorems, definitions, etc., the current custom is to use an elaborate decimal system, such as 2.3.5, meaning item 5 in Section 3 of Chapter 2. I must admit that so many numbers appears repulsive to me, and I assume the reader shares my feelings (perhaps subconsciously). Consequently, I use only positive integers in numbering items. However, in referring to items in other chapters there is some efficiency in a decimal notation. Thus Problem 4.7 means Problem 7 in Chapter 4, and Chapter 5.3 means Section 3 of Chapter 5, etc.

Space does not permit including the proof of every theorem stated. To make matters clear we use the label **PLE** on a theorem to indicate that the "Proof is Left as an Exercise." Similarly, **PWO** means that the "Proof Will be Omitted." In this latter case, we regard the proof as too long or too difficult to be casually treated as an exercise.

We also use a star (★) to indicate that a problem is somewhat difficult.

3. AN APOLOGY

Since I have available only one-half of the more than 4,000 reprints on univalent and multivalent functions, it follows trivially that this book must be full of errors: priority given to the wrong mathematician, important contributions overlooked, etc. Further, I could not include in this book all of the material covered by the 2,000 reprints that I do have. It is my hope that any reader who observes an error, an omission, or a misprint will take the trouble to send me the necessary correction. If all goes well, there may be a second edition in which I will have an opportunity to revise the book and eliminate those errors that have come to light.

4. OMISSIONS

In an effort to cut down the size of this book, I have deliberately omitted a number of topics. The two most important items are (A) Loewner's theory, and (B) the Grunsky inequalites. This loss is not fatal, because these two topics are covered in depth in the excellent book by Peter Duren.

5. THE HISTORY OF THIS BOOK

During the third quarter of the academic year 1975–1976, I was an exchange professor at the University of Paris VII, and this book began as a series of lectures given at that school. Thus I am deeply indebted to the administration of the University of South Florida for arranging this exchange and to the University of Paris VII for the warm reception I was accorded there. The book languished in limbo for several years during which I burdened myself with trivial matters. Again the administration of the University of South Florida came to the rescue by granting me a sabbatical leave for the academic year 1980–1981, and it was during this period that most of the book was written.

6. A ONE-QUARTER COURSE ON UNIVALENT FUNCTIONS

A teacher who selects this book as the text for a one-quarter course should omit certain sections in the early chapters and all of the later chapters.

From the first 11 chapters, the teacher should omit: Chapter 1, all sections; Chapter 5, all sections; Chapter 7, Sections 6, 7, and 8; Chapter 8, Section 10; Chapter 9, Sections 4, 5, and 6; Chapter 10, Sections 5 and 6; Chapter 11, Sections 2 and 3. If there is any time left, the teacher can select at random from the remaining material.

7. ACKNOWLEDGMENTS

I have had the help and friendly advice of many mathematicians: Roger W. Barnard, S. D. Bernardi, Johnny E. Brown, Peter Dörfler, W. E. Kirwan, Richard Libera, Ed Merkes, Sanford S. Miller, Bernard Pinchuk,

Franz Schnitzer, Glenn Schober, and S. M. Shah. It is my pleasure to thank each of them here. A special word of appreciation must go to Douglas M. Campbell, who carefully read Chapters 1 through 17 and made many valuable suggestions for improving the text. I must *emphasize that none of the above-named helpers ever saw Chapter 18 before the book was published and they had absolutely no connection with the contents of that chapter.*

However, I must personally take the blame for any defects this book may have.

Finally, I owe a great debt to my wife, Betty, who cheerfully and accurately typed the entire manuscript. Further, Betty maintained in the household a reasonable degree of quiet, so that my scattered thoughts could be organized into this book.

A. W. Goodman
Tampa, Florida

CONTENTS

Volume I

CONTENTS

Volume II

CHAPTER 1

THE INITIAL CONDITIONS, A REVIEW OF SOME FACTS

We assume that the reader has covered the material usually contained in the first course in complex variables. Here we list a selection of some of the main items together with a few formulas that may be needed in this book.

Each item should be treated by the reader as: (a) something the reader already knows, or (b) something the reader should locate in another text, or (c) something the reader should prove as an exercise.

1. ALGEBRA AND TRIGONOMETRY

An algebraic identity in $z = x + iy$ or $z = re^{i\theta}$ gives further identities when split into real and imaginary parts. As a simple example the identity

$$\frac{1 - z^{n+1}}{1 - z} = 1 + z + z^2 + \cdots + z^n,$$

(true when $z \neq 1$) will give the two identities

(1) $$1 + \sum_{k=1}^{n} r^k \cos k\theta = \frac{1 - r\cos\theta - r^{n+1}\cos(n+1)\theta + r^{n+2}\cos n\theta}{1 - 2r\cos\theta + r^2}$$

and

(2) $$\sum_{k=1}^{n} r^k \sin k\theta = \frac{r\sin\theta - r^{n+1}\sin(n+1)\theta + r^{n+2}\sin n\theta}{1 - 2r\cos\theta + r^2}.$$

We can replace r by $z = \varrho e^{i\phi}$ in (1) and (2) and repeat the process of splitting into real and imaginary parts, but the results are more complicated and seem to be less useful. If we let $n \to \infty$ in (1) and (2) and assume $r < 1$ so that the series converges, we obtain

(3) $$1 + \sum_{k=1}^{\infty} r^k \cos k\theta = \frac{1 - r\cos\theta}{1 - 2r\cos\theta + r^2}$$

and

(4) $$\sum_{k=1}^{\infty} r^k \sin k\theta = \frac{r\sin\theta}{1 - 2r\cos\theta + r^2}.$$

We can also keep n finite and set $r = 1$ in (1) and (2). After a few trigonometric transformations we will find that

(5) $$1 + \sum_{k=1}^{n} \cos k\theta = \frac{\sin[(n+1)\theta/2]}{\sin(\theta/2)} \cos\frac{n\theta}{2},$$

and

(6) $$\sum_{k=1}^{n} \sin k\theta = \frac{\sin[(n+1)\theta/2]}{\sin(\theta/2)} \sin\frac{n\theta}{2}.$$

We list a few other identities that can be derived either by the previous method or by some other method such as mathematical induction. We leave it for the reader to consider the situation when a denominator is zero or to discuss the region of convergence of any infinite series.

(7) $$1 + 2\sum_{k=1}^{\infty} r^k \cos k\theta = \frac{1 - r^2}{1 - 2r\cos\theta + r^2}.$$

(8) $$\sum_{k=1}^{\infty} k r^k \sin k\theta = r\sin\theta \frac{1 - r^2}{(1 - 2r\cos\theta + r^2)^2}.$$

(9) $$\sum_{k=1}^{n} \cos(2k-1)\theta = \frac{\sin 2n\theta}{2\sin\theta}.$$

(10) $$\sum_{k=1}^{n} \sin(2k-1)\theta = \frac{\sin^2 n\theta}{\sin\theta}.$$

(11) $$\frac{n+1}{2} + \sum_{k=1}^{n} (n+1-k)\cos k\theta = \frac{\sin^2[(n+1)\theta/2]}{2\sin^2(\theta/2)}.$$

(12) $$\sum_{k=1}^{n} (n+1-k)\sin k\theta = \frac{(n+1)\sin\theta - \sin(n+1)\theta}{4\sin^2(\theta/2)}$$

The function in (7) is the Poisson kernel.

The trigonometric functions are orthogonal. If k and m are *distinct* integers,

(13) $$\int_0^{2\pi} \sin k\theta \sin m\theta \, d\theta = \int_0^{2\pi} \cos k\theta \cos m\theta \, d\theta = 0.$$

For all integers k, m

(14) $$\int_0^{2\pi} \sin k\theta \cos m\theta \, d\theta = 0.$$

Of course, if $k \neq 0$,

(15) $$\int_0^{2\pi} \sin^2 k\theta \, d\theta = \int_0^{2\pi} \cos^2 k\theta = \pi.$$

(16) $$\int_0^{2\pi} e^{ik\theta} e^{im\theta} \, d\theta = \begin{cases} 0, & \text{if } k+m \neq 0, \\ 2\pi, & \text{if } k+m = 0. \end{cases}$$

If $\eta = e^{2\pi i/N}$ where N is a positive integer and k is any integer, then

(17) $$\sum_{j=0}^{N-1} (\eta^j)^k = \begin{cases} 0 \text{ if } k \not\equiv 0 \pmod{N}, \\ N \text{ if } k \equiv 0 \pmod{N}. \end{cases}$$

For any q the binomial theorem states that

(18) $$\frac{1}{(1-z)^q} = 1 + qz + \sum_{n=2}^{\infty} \frac{q(q+1)\cdots(q+n-1)}{n!} z^n,$$

Of course, if $q \leq 0$ is an integer, then (18) is a polynomial.

The Gamma function $\Gamma(x)$ is defined by

(19) $$\Gamma(x) = \int_0^{\infty} e^{-t} t^{x-1} \, dt, \qquad x > 0.$$

When $n \geq 0$ is an integer, $\Gamma(n+1) = n!$. To approximate $n!$ when n is large, we have Stirlings formula

$$n! \sim \sqrt{2\pi n}\left(\frac{n}{e}\right)^n. \tag{20}$$

The function $\sin \pi z$ is given by an infinite product

$$\sin \pi z = \pi z \prod_{n=1}^{\infty}\left(1 - \frac{z^2}{n^2}\right). \tag{21}$$

If we put $z = 1/2$, we obtain the Wallis formula for π,

$$\frac{\pi}{2} = \lim_{n\to\infty} \frac{2\cdot 2\cdot 4\cdot 4\cdot 6\cdot 6 \cdots (2n)\cdot(2n)}{1\cdot 3\cdot 3\cdot 5\cdot 5\cdot 7 \cdots (2n-1)(2n+1)}. \tag{22}$$

2. SOME INEQUALITIES

The arithmetic, geometric, harmonic inequality asserts that for any finite sequence of positive numbers $a_1, \ldots, a_n$

$$A \geq G \geq H \tag{23}$$

where

$$A = \frac{1}{n}\sum_{k=1}^{n} a_k, \qquad G = \left(\prod_{k=1}^{n} a_k\right)^{1/n}, \qquad H = \frac{n}{\sum_{k=1}^{n} 1/a_k}.$$

We can replace the sums by integrals. If $f(x)$ has a positive lower bound and is integrable in $[a, b]$, then (23) still holds where now

$$A = \frac{1}{b-a}\int_a^b f(x)\,dx, \qquad G = \exp\left(\frac{1}{b-a}\int_a^b \ln f(x)\,dx\right),$$

$$H = \frac{b-a}{\int_a^b (1/f(x))\,dx}.$$

The Cauchy-Bunyakowsky-Schwarz (CBS) inequality states that for any two sequences of numbers

$$\left|\sum_{k=1}^{n} a_k b_k\right|^2 \leq \left(\sum_{k=1}^{n} |a_k|^2\right)\left(\sum_{k=1}^{n} |b_k|^2\right) \tag{24}$$

with equality if and only if the sets are proportional. If $f(x)$ and $g(x)$ are integrable in $[a, b]$, then (24) can be extended to give

$$\left|\int_a^b f(x)g(x)\,dx\right|^2 \le \left(\int_a^b |f(x)|^2\,dx\right)\left(\int_a^b |g(x)|^2\,dx\right). \tag{25}$$

Suppose that $f(x)$ and $g(x)$ are nonnegative and integrable in $[a, b]$ and p and q are positive numbers for which

$$\frac{1}{p} + \frac{1}{q} = 1. \tag{26}$$

Hölder's inequality asserts that under these conditions

$$\int_a^b f(x)g(x)\,dx \le \left(\int_a^b [f(x)]^p\,dx\right)^{1/p}\left(\int_a^b [g(x)]^q\,dx\right)^{1/q}. \tag{27}$$

Equality occurs if and only if $[f(x)]^p = C[g(x)]^q$ almost everywhere. The CBS inequality is a special case of (27) when $p = q = 2$. Exponents p and q that satisfy (26) are called *conjugate exponents*. The condition (26) can be put in the form $(p-1)(q-1) = 1$ or $p = q/(q-1)$.

Suppose that $0 < q \le 1$. Then Minkowski's inequality states that

$$\left|\sum_{k=1}^n a_k^q\right|^{1/q} + \left|\sum_{k=1}^n b_k^q\right|^{1/q} \le \left(\sum_{k=1}^n (|a_k| + |b_k|)^q\right)^{1/q}. \tag{28}$$

If $f(x)$ and $g(x)$ are nonnegative and integrable in $[a, b]$, and $0 < q \le 1$, then the integral form of Minkowski's inequality is

$$\left(\int_a^b [f(x)]^q\,dx\right)^{1/q} + \left(\int_a^b [g(x)]^q\,dx\right)^{1/q} \le \left(\int_a^b (f(x)+g(x))^q\,dx\right)^{1/q}. \tag{29}$$

If $q > 1$, then the inequality sign is reversed in (28) and (29).

3. DIFFERENTIATION

Let $f(z) = u(x, y) + iv(x, y)$ where $z = x + iy$. A necessary and sufficient condition for $f(z)$ to be regular (analytic) in a domain D is that $f(z)$ is continuous in D and

$$u_x = v_y \quad \text{and} \quad u_y = -v_x \tag{30}$$

in D. These are the *Cauchy-Riemann* equations.

We may also introduce polar coordinates by using $z = re^{i\theta}$ in D, or $w = Re^{i\Phi}$ in the image domain $f(D)$. Thus the transformation $w = f(z)$ can be regarded in four different ways with respect to the coordinate systems used: (I) rect. → rect., (II) rect. → polar, (III) polar → rect., and (IV) polar → polar. This gives rise to four different sets of Cauchy-Riemann equations. Thus (30) is the set for (I). The other three are:

$$R_x = R\Phi_y, \quad \text{and} \quad R_y = -R\Phi_x, \quad \text{when } f(x+iy) = Re^{i\Phi}, \tag{31}$$

$$ru_r = v_\theta, \quad \text{and} \quad u_\theta = -rv_r, \quad \text{when } f(re^{i\theta}) = u+iv, \tag{32}$$

$$rR_r = R\Phi_\theta, \quad \text{and} \quad R_\theta = -Rr\Phi_r, \quad \text{when } f(re^{i\theta}) = Re^{i\Phi}. \tag{33}$$

For example, to prove (31) one may start with the analytic function[1] $\ln f(z) = \ln R + i\Phi = u + iv$ and use (30).

When $w = f(z)$ is regarded as a transformation from a polar coordinate domain to a rectangular coordinate domain, Fejér and Szegö 1951 [1-B-301] give a large number of interesting formulas. As one simple example, if $f(z)$ is analytic, then $u_\theta + iv_\theta = izf'(z)$.

Let $f(z) = u(x, y) + iv(x, y)$ where $z = x + iy$. It is convenient to define partial derivatives f_z and $f_{\bar z}$ by

$$f_z \equiv \frac{\partial f}{\partial z} \equiv \frac{1}{2}(f_x - if_y) = \frac{1}{2}(u_x + iv_x - iu_y + v_y) \tag{34}$$

and

$$f_{\bar z} \equiv \frac{\partial f}{\partial \bar z} \equiv \frac{1}{2}(f_x + if_y) = \frac{1}{2}(u_x + iv_x + iu_y - v_y). \tag{35}$$

Here we assume that the partial derivatives exist in some domain D. These operators f_z and $f_{\bar z}$ are often useful. If $f(z)$ is regular in D, then

$$f_{\bar z} = 0 \quad \text{and} \quad f_z = f'(z). \tag{36}$$

Conversely, if either $f_{\bar z} = 0$ or $f_z = f'(z)$ in D, then $f(z)$ is regular in D. Further, we have

$$|f_z|^2 - |f_{\bar z}|^2 = J\left(\frac{u, v}{x, y}\right) \equiv \begin{vmatrix} u_x & u_y \\ v_x & v_y \end{vmatrix}. \tag{37}$$

[1] Of course, we are assuming that $f(z) \neq 0$.

Suppose that $f(z)$ is regular in a domain D, and z is some differentiable function of a *real variable t*. Then we can apply the chain rule and obtain

$$\frac{df(z)}{dt} = f'(z)\frac{dz}{dt}. \tag{38}$$

In fact the classical formal proof

$$\frac{df}{dt} = \lim_{\Delta t \to 0} \frac{\Delta f}{\Delta t} = \lim_{\Delta t \to 0} \frac{\Delta f}{\Delta z}\frac{\Delta z}{\Delta t}$$

can be made rigorous quite easily. Further, because t is a *real variable*, we can write

$$\frac{d}{dt}\operatorname{Re} f(z) = \operatorname{Re}\frac{df}{dt} = \operatorname{Re}\left(f'(z)\frac{dz}{dt}\right). \tag{39}$$

Equation (39) is very useful when $z = re^{it}$ (z describes a circle) and when $z = te^{i\alpha}$ (z describes a radial line).

As an example, if $z = re^{it}$, then consideration of

$$\ln |f(z)| = \operatorname{Re} \ln f(z)$$

gives

$$\frac{d}{dt}|f(z)| = -|f(z)| \operatorname{Im} \frac{zf'(z)}{f(z)}. \tag{40}$$

4. INTEGRATION

Throughout this section Γ is a simple closed piecewise smooth curve and (unless otherwise stated) $f(z)$ is regular on Γ and in the domain D enclosed by Γ. The point z is in D and ζ is on Γ.

Cauchy's Theorem. Under the above conditions

$$\int_\Gamma f(\zeta)\, d\zeta = 0. \tag{41}$$

In his proof Cauchy assumed that $f'(z)$ is continuous on $\Gamma \cup D$. Goursat removed this additional hypothesis, and hence (41) is sometimes called the *Cauchy-Goursat Theorem*.

Cauchy's Integral Formula. For $n = 0, 1, 2, \ldots$

(42) $$f^{(n)}(z) = \frac{n!}{2\pi i} \int_{\Gamma} \frac{f(\zeta)}{(\zeta - z)^{n+1}} \, d\zeta.$$

If $f(z) = \Sigma_0^\infty a_n z^n$ for $|z| < R$, then for $n = 0, 1, 2, \ldots$ and $0 < r < R$

(43) $$a_n = \frac{1}{2\pi i} \int_{C_r} \frac{f(\zeta)}{\zeta^{n+1}} \, d\zeta = \frac{1}{2\pi} \int_0^{2\pi} \frac{f(re^{i\theta})e^{-in\theta}}{r^n} \, d\theta.$$

More generally, suppose that $f(z)$ is regular in the ring domain $a < |z| < b$ and has the Laurent expansion

(44) $$f(z) = \sum_{n=-\infty}^{\infty} a_n z^n.$$

Then for each r in (a, b) and each integer n, equation (43) gives a_n.

Poisson's Formula. Let $u(re^{i\theta})$ be harmonic in the closed disk $r \leq R$. Then for each r in $(0, R)$

(45) $$u(re^{i\theta}) = \frac{R^2 - r^2}{2\pi} \int_0^{2\pi} \frac{u(Re^{i\phi}) \, d\phi}{R^2 - 2Rr\cos(\theta - \phi) + r^2}.$$

More generally if $u(Re^{i\phi})$ is any piecewise continuous real-valued function on the circle C_R, then (45) defines a function that is harmonic in the open disk $r < R$ and continuous in the closed disk except for the points of discontinuity of the given function $u(Re^{i\phi})$. This means that $u(re^{i\theta}) \to u(Re^{i\theta})$ as $r \to R^-$ whenever $Re^{i\theta}$ is a point of continuity of $u(Re^{i\phi})$.

In the next five formulas we suppose that u and v are real-valued functions of x and y and that all the quantities involved are continuous in $\bar{D}$.

Green's Theorem in the plan asserts that

(46) $$\int_{\Gamma} u \, dx + v \, dy = \iint_D (v_x - u_y) \, dx \, dy.$$

As a special case of (46) we have

(47) $$\text{Area } (D) = \frac{1}{2} \int_{\Gamma} (x \, dy - y \, dx).$$

Let $\partial u/\partial n$ denote the directional derivative in the direction of the outward normal to the curve Γ, and let $\nabla^2 u = u_{xx} + u_{yy}$. If u is harmonic in D, then $\nabla^2 u \equiv 0$. With this notation:

(48) $$\int_\Gamma \frac{\partial u}{\partial n}\, ds = \iint_D \nabla^2 u \, dx\, dy,$$

where s is arc length on Γ.

(49) $$\int_\Gamma u \frac{\partial v}{\partial n}\, ds = \iint_D (u\nabla^2 v + u_x v_x + u_y v_y)\, dx\, dy.$$

(50) $$\int_\Gamma \left(u \frac{\partial v}{\partial n} - v \frac{\partial u}{\partial n}\right) ds = \iint_D (u\nabla^2 v - v\nabla^2 u)\, dx\, dy.$$

Integral formulas involving complex-valued integrands can be proved by taking real and imaginary parts and using formulas (46) through (50). If the quantities involved are continuous in $\bar{D}$, then

(51) $$\frac{1}{2i}\int_\Gamma \phi \, dz = \iint_D \frac{\partial \phi}{\partial \bar{z}}\, dx\, dy,$$

and

(52) $$-\frac{1}{2i}\int_\Gamma \phi \, d\bar{z} = \iint_D \frac{\partial \phi}{\partial z}\, dx\, dy.$$

If we set $\phi = pq$, then (51) and (52) give

(53) $$\frac{1}{2i}\int_\Gamma pq\, dz = \iint_D \left(p\frac{\partial q}{\partial \bar{z}} + q\frac{\partial p}{\partial \bar{z}}\right) dx\, dy$$

and

(54) $$-\frac{1}{2i}\int_\Gamma pq\, d\bar{z} = \iint_D \left(p\frac{\partial q}{\partial z} + q\frac{\partial p}{\partial z}\right) dx\, dy.$$

If $f(z)$ and $g(z)$ are regular in $\bar{D}$, set $p = f$ and $q = \bar{g}$ in (53). Using $f_{\bar{z}} = 0$, (53) gives

(55) $$\frac{1}{2i}\int_\Gamma f\bar{g}\, dz = \iint_D f\overline{g'}\, dx\, dy.$$

Similarly, if $f(z)$ and $g(z)$ are regular in $\bar{D}$, equation (54) gives

$$-\frac{1}{2i}\int_{\Gamma} f\bar{g}\, d\bar{z} = \iint_{D} \bar{g}f'\; dx\, dy. \tag{56}$$

If we set $g(z) \equiv z$ in (55), we obtain

$$\frac{1}{2i}\int_{\Gamma} \bar{z}f(z)\; dz = \iint_{D} f(z)\; dx\, dy. \tag{57}$$

If we set $f(z) = g'(z)$ in (55), we find that

$$\frac{1}{2i}\int_{\Gamma} g'(z)\overline{g(z)}\; dz = \iint_{D} |g'(z)|^2 dx\, dy. \tag{58}$$

All of the above formulas can be extended to multiply-connected domains, when suitable agreements are made about Γ.

5. SOME GEOMETRIC FUNCTION THEORY

If $|a| < 1$ and α is a real number, then the function

$$\phi(z) = e^{i\alpha}\frac{z-a}{1-\bar{a}z} \tag{59}$$

maps E conformally onto itself. Conversely, any function ϕ which maps E conformally onto itself must have the form (59).

The set of all function of the form (59) form a group. This group is a subgroup of the group of all linear (Möbius) functions

$$\phi(z) = \frac{Az+B}{Cz+D}, \qquad AD-BC \neq 0. \tag{60}$$

Each function of the form (60) is meromorphic in the closed complex plane (includes the point at ∞) and maps the closed complex plane in a one-to-one manner onto itself. Conversely, any $\phi(z)$ with these properties must have the form (60). A function of the form (60) always maps a circle or line onto a circle or line, and any circular domain or half-plane onto a circular domain or half-plane.

The Riemann Mapping Theorem. Let D be a simply-connected domain in the closed complex plane with at least two boundary points, and let z_0 be a fixed point in D. Then there is one and only one function $f(z)$ meromorphic[1] in E that maps E in a one-to-one manner onto D with $f(0) = z_0$ and $f'(0) > 0$.

Examples illustrating this theorem are sprinkled throughout this text. It we are permitted to shrink or expand D, then we can always find a domain $\tilde{D}$ similar to D for which $f'(0) = 1$.

[1] We say "meromorphic" rather than "regular" because $w = \infty$ may be an interior point of D.

CHAPTER 2

THE DEFINITION AND ELEMENTARY PROPERTIES OF UNIVALENT FUNCTIONS

1. SOME DEFINITIONS AND EXAMPLES

A function $f(z)$ is said to be univalent in a domain D if it provides a one-to-one mapping onto its image, $f(D)$. Geometrically, this means that the representation of the image domain can be visualized as a suitable set of points in the complex plane. This rough intuitive description is made precise in

Definition 1. A function $f(z)$ is said to be *univalent in a domain D* if the conditions

$$f(z_1) = f(z_2), \qquad z_1 \in D, \qquad z_2 \in D, \tag{1}$$

imply that $z_1 = z_2$.

Various other terms are used for this concept. Thus univalent functions are called *simple*, or *schlicht* (the German word for simple). The

Russians refer to such functions as *odnolistni*, which means single-sheeted.

We are mainly interested in univalent functions that are also regular (analytic, holomorphic) in D. Consequently, the term univalent carries with it the connotation that $f(z)$ is also regular. But we do not include this item in the definition because on many occasions we well permit $f(z)$ to have a simple pole in D (if $f(z)$ is univalent, it cannot have a pole of higher order, or more than one pole in D).

The theory of univalent functions is so vast and complicated that certain simplifying assumptions are necessary. The most obvious one is to replace the arbitrary domain D by one that is convenient, and the most attractive selection is the unit disk E: $|z| < 1$.

As an elementary example consider the function $f(z) = (1+z)^2$ in the unit disk E. The univalence of this function in E is easy to see on geometric grounds. Indeed $1+z$ shifts the unit disk to the right and the effect of squaring $1+z$ is easy to visualize. The three domains are shown in Figure 1. (See next page.)

The same type of argument shows that $w = (1+z)^3$ is *not* univalent in E.

If $g(z)$ is regular in E, it has a Maclaurin expansion

$$g(z) = b_0 + b_1 z + b_2 z^2 + \cdots = \sum_{n=0}^{\infty} b_n z^n, \tag{2}$$

that is convergent in E. It is natural to ask two questions about this representation of $g(z)$. Given the sequence of coefficients $\{b_n\}$, how does this sequence influence the geometric properties of $g(z)$? Given some property of $g(z)$, how does this property influence the coefficients in (2)? We are concerned with the property of univalence, so these two questions specialize thus: (I) If $g(z)$ is a univalent function, how does this fact restrict the coefficients in the Maclaurin series for $g(z)$? (II) Given the coefficients in (2), can we decide if $g(z)$ is univalent in D?

We observe that if $g(z)$ is univalent in E then the addition of a constant merely translates the image domain so that $g(z)+C$ is again univalent in E. Consequently in (2), the constant term b_0 is arbitrary. As a first step toward normalization we subtract b_0, and consider $g(z) - b_0$. We next observe that if $g'(z_0) = 0$, then $g(z)$ is not univalent in any neighborhood of z_0. Consequently if $g(z)$ is univalent in E, then $b_1 = g'(0) \neq 0$. Hence we may divide by b_1 and consider $f(z) \equiv (g(z) - b_0)/b_1$.

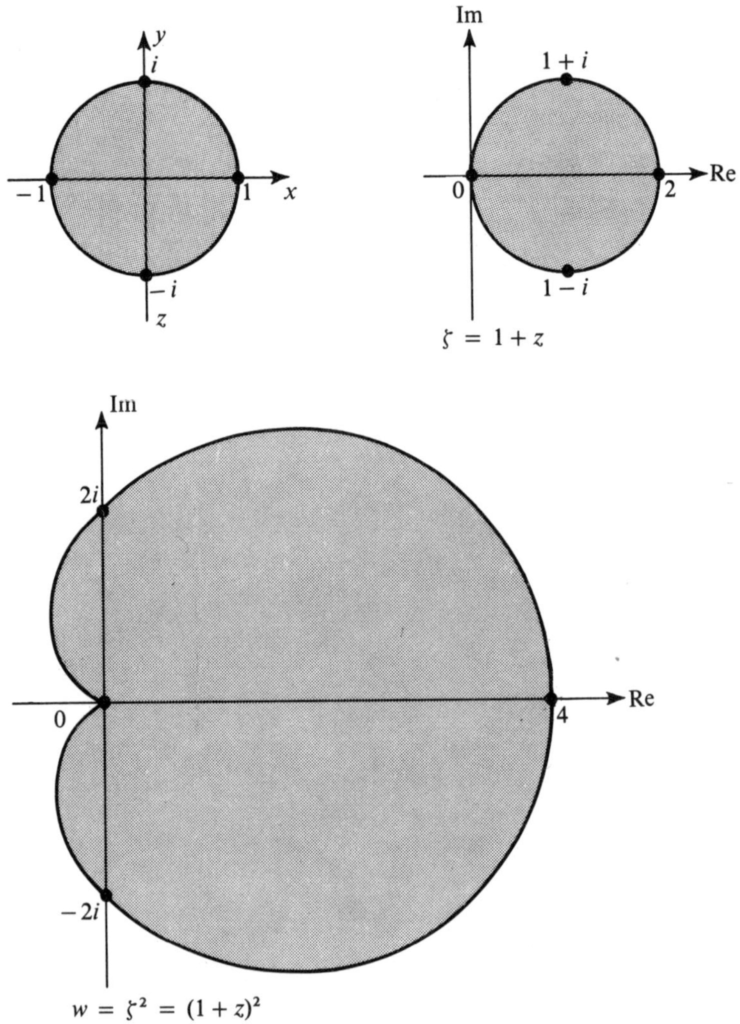

FIGURE 1.

Since multiplication by $1/b_1$ merely rotates and stretches (or shrinks) the image domain, we see that if $g(z)$ is univalent in D then $f(z) \equiv (g(z) - b_0)/b_1$ is also univalent in the same domain and, conversely, if $f(z)$ is univalent in D, then so also is $g(z)$. Setting $b_n/b_1 = a_n$ in (2) we arrive at the "normalized" form

$$f(z) = z + a_2z^2 + a_3z^3 + \cdots = z + \sum_{n=2}^{\infty} a_nz^n. \tag{3}$$

Definition 2. A function of the form (3) is said to be *normalized.* If $f(z)$ is univalent and has the form (3), it is called a *normalized univalent function.* The class[1] of all normalized functions that are regular, and univalent in E, is denoted by S (simple, schlicht[2]).

It is clear from the discussion that if $g(z)$, given by (2) is univalent in E, then $|b_n|$ may be arbitrarily large (multiply $g(z)$ by a suitable M). However, if we consider the class S of normalized univalent functions, we might conjecture that for each $n > 2$ there is a bound A_n such that $|a_n| < A_n$ for every $f(z)$ in S. To obtain some estimate for A_n we proceed to construct a univalent function that has "large" coefficients. Indeed, consider the sequence

$$u(z) = \frac{1+z}{1-z}, \qquad g(z) = u^2(z), \qquad f(z) = \frac{1}{4}(g(z) - 1). \tag{4}$$

The function $u(z)$ carries E onto the half-plane $\operatorname{Re} u > 0$. Then $g(z) = u^2(z)$ takes this half-plane onto the entire complex plane except for the negative real axis: $g \le 0$. Finally $f(z) = (g(z) - 1)/4$ is merely the normalization process described above. To compute $f(z)$ we have

$$f(z) = \frac{1}{4}\left[\left(\frac{1+z}{1-z}\right)^2 - 1\right] = \frac{(1+z)^2 - (1-z)^2}{4(1-z)^2} = \frac{z}{(1-z)^2}.$$

This function is so important that it has a special name, *the Koebe function,* and a special symbol, $k(z)$. The sequence of mappings used in building $k(z)$ is indicated in Figure 2. (See next page.)

Theorem 1. The Koebe function

$$k(z) \equiv \frac{z}{(1-z)^2} = z + 2z^2 + 3z^3 + \cdots + nz^n + \cdots = \sum_{n=1}^{\infty} nz^n \tag{5}$$

[1] In this book we use the words "class," "family," and "set" to mean the same thing, namely, the collection of all functions that have the property under consideration.

[2] Other symbols, such as U and $V(1)$, occur in the literature. Of course U denotes univalent. The symbol $V(p)$ means the class of functions with valence p. However, S seems to have won universal approval.

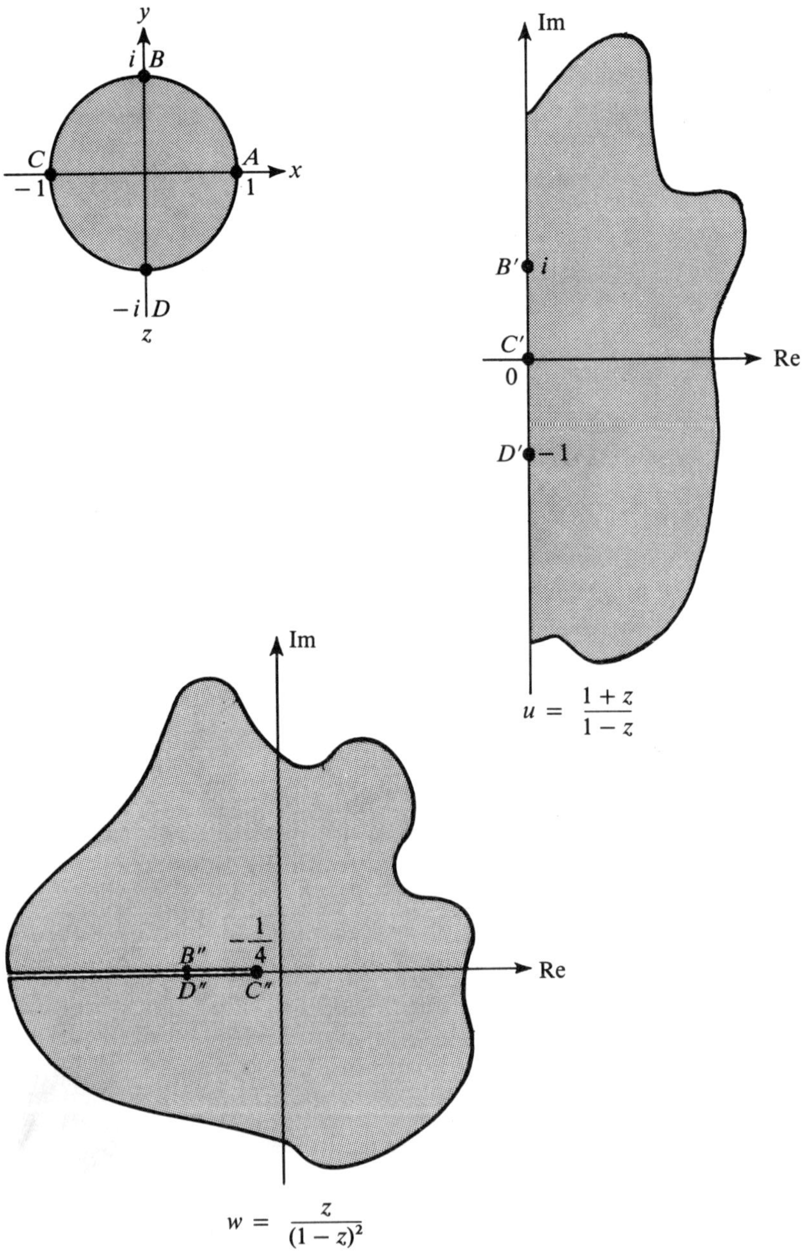
y
i B
C
A
x
−1
1
−i D
z
Im
B′ i
C′
Re
0
D′ −1
$u = \frac{1+z}{1-z}$
Im
$-\frac{1}{4}$
B″
D″
C″
Re
$w = \frac{z}{(1-z)^2}$

FIGURE 2.

is in S. It maps E in a one-to-one manner onto the domain D that consists of the entire complex plane except for a slit along the negative real axis from $w = -\infty$ to $w = -1/4$.

In an intuitive sense this function is the "largest" function in S, because it is impossible to add to the image domain any open set of points without destroying univalence. However, this is not the only domain with this property. If we let Γ be any curve from the point at infinity to some finite w_0, then there is a function that is regular and univalent in E and maps E onto the complement of Γ. Nevertheless, the "maximal" nature of the domain in Theorem 1, its symmetry, and the size of the coefficients of $k(z)$ suggest the

Conjecture. If $f(z)$ in S has the Maclaurin series (3), then for each $n \geq 2$

$$|a_n| \leq n. \tag{6}$$

This conjecture is often referred to in the literature as the Bieberbach conjecture, because he stated it in a footnote in 1916 [1-B-117, p. 946]. However, because of the extremal properties of the Koebe function, it is an easy conjecture to make, and in fact many mathematicians made this same conjecture during the second decade of this century. In this text we will refer to this conjecture as the "central conjecture" of the theory of univalent functions, or as the "coefficient conjecture."

At this time the conjecture is still open although it has been proved in many special cases, some of which will be covered in later chapters. The inequality $|a_n| \leq n$ has been proved for $n = 2, 3, 4, 5$, and 6, but at present it is open for all $n \geq 7$. D. Horowitz 1978 [3-B-368], using a method due to FitzGerald 1972 [2-B-395], has proved that $|a_n| < 1.0657n$ for all n. For a history of results of this type together with references see D. Horowitz [3-B-368].

2. SOME OPERATIONS IN THE SET OF UNIVALENT FUNCTIONS

In a systematic investigation of any class of functions, it is helpful to have operations that take a function of the set into another function of the same set. We collect four of these elementary operations (together with their series expansions) for the class S in

Theorem 2. PLE. If $f(z)$, given by (3), is in S then (under the specified conditions) each of the following functions is also in S:

$$e^{-i\alpha}f(e^{i\alpha}z) = z + \sum_{n=2}^{\infty} a_n e^{i(n-1)\alpha} z^n, \qquad \alpha \text{ real}, \tag{7}$$

$$\overline{f(\bar{z})} = z + \sum_{n=2}^{\infty} \bar{a}_n z^n, \tag{8}$$

$$\frac{1}{t} f(tz) = z + \sum_{n=2}^{\infty} a_n t^{n-1} z^n, \qquad 0 < t \le 1, \tag{9}$$

$$[f(z^k)]^{1/k} = z + \frac{a_2}{k} z^{k+1} + \frac{1}{2k^2}(2ka_3 - (k-1)a_2^2)z^{2k+1} + \cdots, \tag{10}$$

where k is a positive integer.

Henceforth we will speak of the transformation (7) as a "rotation of $f(z)$," because it rotates the unit disk in the z-plane through an angle α and rotates the image domain in the w-plane in the reverse direction through an angle of the same magnitude. We include $f(z)$ itself in the term "rotation of $f(z)$," because we can always set $\alpha = 0$ in (7).

In (10) the transformation $u = z^k$ maps E onto k copies of E, and $f(u)$ carries this surface onto k copies of $f(E)$ joined by a suitable branch point at $w = 0$. It is intuitively clear that the kth root merely "unwinds" the surface to form a plane domain with k-fold symmetry. Precisely, we have

Definition 3. Let k be a positive integer. A domain D is said to be *k-fold symmetric* if a rotation of D about the origin through an angle $2\pi/k$ carries D onto itself. A function $f(z)$ is said to be k-fold symmetric in E if for every z in E

$$f(e^{2\pi i/k}z) = e^{2\pi i/k} f(z). \tag{11}$$

Theorem 3. PLE. Gronwall 1916 [1-B-454]. If $g(z)$ is regular and k-fold symmetric in E, then it has a power series expansion of the form

$$g(z) = b_1 z + b_{k+1} z^{k+1} + b_{2k+1} z^{2k+1} + \cdots = \sum_{n=0}^{\infty} b_{nk+1} z^{nk+1}. \tag{12}$$

Conversely, if $g(z)$ is given by the power series (12), then $g(z)$ is k-fold symmetric inside the circle of convergence of the series.

Note that in Definition 3 and Theorem 3 we do not require that $f(z)$ be univalent. Definition 3 can be applied to a domain D on a Riemann surface if suitable agreements are made. Our interest centers on normalized univalent functions, and we will use $S^{(k)}$ to denote the subclass of functions in S that are also k-fold symmetric. If $f(z)$ is in $S^{(2)}$, then $f(z)$ is called an *odd univalent function*.

Two more transformations that carry S onto S are covered in the next two theorems.

Theorem 4. PLE. If $f(z)$ is in S and

(13) $$\phi(z) = \frac{z+\lambda}{1+\bar{\lambda}z}, \qquad |\lambda| < 1,$$

then

(14) $$g(z) \equiv C[f(\phi(z)) - f(\lambda)] = z + b_2 z^2 + \cdots$$

is also in S, where the normalizing factor C is given by

(15) $$C = \frac{1}{f'(\lambda)\phi'(0)} = \frac{1}{f'(\lambda)(1-|\lambda|^2)},$$

and

(16) $$b_2 = \frac{(1-|\lambda|^2)f''(\lambda)}{2f'(\lambda)} - \bar{\lambda}.$$

Theorem 5. PLE. Suppose that $f(z)$ is in S and $f(z)$ omits[1] γ in E. Then

(17) $$g(z) = \frac{f(z)}{1-f(z)/\gamma} = z + (a_2 + 1/\gamma)z^2 + \cdots$$

is also in S.

In due time we will find that it is often useful to replace the disk E by

[1] We say that a function $f(z)$ omits γ in a domain D if the equation $f(z) = \gamma$ has no solution with z in D.

the exterior of the unit circle, and the function $f(z)$ by its reciprocal $1/f(z)$. In anticipation of these operations we introduce more symbols:

E^x denotes the domain $1 < |\zeta| < \infty$.

Σ is the class of all functions

$$\phi(\zeta) = \zeta + c_0 + \frac{c_1}{\zeta} + \frac{c_2}{\zeta^2} + \cdots = \zeta + \sum_{n=0}^{\infty} \frac{c_n}{\zeta^n} \tag{18}$$

that are regular and univalent in E^x. The subclass of those functions that omit $\gamma = 0$ will be denoted by Σ_0.

Theorem 6. PLE. If $f(z)$, given by (3), is in S, then

$$\phi(\zeta) \equiv \frac{1}{f(1/\zeta)} = \zeta - a_2 + (a_2^2 - a_3)\frac{1}{\zeta} + \cdots \tag{19}$$

is in Σ_0. Conversely, if $\phi(\zeta)$, given by (18), is in Σ_0, then

$$\begin{aligned} f(z) &= \frac{1}{\phi(1/z)} \\ &= z - c_0 z^2 + (c_0^2 - c_1)z^3 + (2c_0c_1 - c_2 - c_0^3)z^4 + \cdots \end{aligned} \tag{20}$$

is in S.

If the transformation (19) is applied to the Koebe function, the resulting $\phi(\zeta)$ takes on a particularly simple form. Indeed, we find that

$$\phi(\zeta) = \frac{1}{k(1/\zeta)} = \zeta - 2 + \frac{1}{\zeta}, \tag{21}$$

a function that maps E^x onto the doubly-connected domain consisting of the entire complex plane minus the slit $-4 \leq w \leq 0$. If this image domain is regarded in the closed complex plane (∞ included), then it is simply-connected.

PROBLEMS

1. Give an algebraic proof (see Definition 1) that $f(z) \equiv (1 + z)^2$ is univalent in E.
2. Let $k \geq 2$ be a fixed positive integer. Find the smallest $A > 0$ such that $(A + z)^k$ is univalent in E.

3. Prove that the function

(22)
$$f(z) \equiv \frac{1}{2} \ln \frac{1+z}{1-z} = z + \frac{z^3}{3} + \frac{z^5}{5} + \cdots + \frac{z^{2n+1}}{2n+1} + \cdots$$

has the power series indicated, and that $f(z)$ is univalent in E. What is $f(E)$ for this function?

4. Let D be a domain in the z-plane with the property that each vertical line ($\text{Re } z =$ constant) meets D in a single line segment of length less than 2π. Prove that e^z is univalent in D.

5. Combine the results of Problems 3 and 4 to show that if we have $-\pi/2 < \alpha < \pi/2$ and $0 < c \le 2 \cos \alpha$, then the function

(23)
$$f(z) \equiv \exp \left\{ ce^{i\alpha} \ln \frac{1+z}{1-z} \right\} \equiv \left(\frac{1+z}{1-z} \right)^{ce^{i\alpha}}$$

is univalent in E. Here $\exp Q \equiv e^Q$.

6. Find necessary and sufficient conditions on a_2 for $f(z) \equiv z + a_2 z^2$ to be univalent in E. What are these conditions for the non-normalized function $g(z) = b_0 + b_1 z + b_2 z^2$? Note that the same problem for a cubic polynomial is not trivial. See Cowling and Royster 1968 [2-B-294], Brannan 1967 [2-B-171], and 1970 [2-B-174]. For the trinomial $f(z) = z + az^k + bz^n$ see Kasten and Schmieder 1980 [3-B-435].

7. Prove that if $f(z)$ has the power series (3) and $\Sigma_{n=2}^{\infty} n|a_n| \le 1$, then $f(z)$ is univalent in E. Is this sufficient condition for univalence in E also necessary? Hurwitz 1897 [3-B-379], Remak 1943 [1-B-1124]. The subset of all functions in S that satisfy the condition $\Sigma_{n=2}^{\infty} n|a_n| \le 1$ will be denoted by HZ.

8. By differentiating $1/(1-z) = \Sigma_{n=0}^{\infty} z^n$, show that the Koebe function has the Maclaurin series $\Sigma_{n=1}^{\infty} nz^n$.

9. Give geometric arguments to prove Theorem 2.

10. Describe the image domain $g(E)$ if $g(z)$ is obtained from the Koebe function under the transformation of: Equation (7), Equation (8), and Equation (10).

11. Prove Theorem 3.

12. Prove Theorems 4, 5, and 6.

13. Check that the power series expansions given in Theorems 2, 4, 5, and 6 are correct.

14. Show that the coefficient conjecture (6) is true for the univalent functions found in Problem 2. Hint: Remember to normalize $g(z) \equiv (A + z)^k$.

15. Find a conjecture for non-normalized univalent functions that is equivalent to $|a_n| \leq n$ for functions in S.

16. Find the first three non-zero coefficients of $\phi(\zeta)$ when the transformation of Theorem 6 is applied to $[f(z^k)]^{1/k}$. A formula for all the coefficients is given in equation (5.20).

17. Explain the need for the class Σ_0 (in place of Σ) in Theorem 6. What type of function is obtained if the transformation (20) is applied to a function $\phi(\zeta)$ in Σ?

18. Apply the transformation (20) to the function $\phi(\zeta) = \zeta - b + 1/\zeta$, where $b > 2$, to show that if $b = p + 1/p$, then the function

$$k_p(z) = \frac{z}{1 - bz + z^2} \tag{24}$$

is univalent in E and is regular in E except for the pole at $z = p < 1$. Find $k_p(E)$. What is $k_p(z)$ when $p = 1$?

19. Find a_n for the Maclaurin series for $k_p(z)$ given in Problem 18. Hint: Use partial fractions.

20. Let $S(p)$ denote the set of normalized functions that are univalent and regular in E except for a simple pole at z_0, where $|z_0| = p < 1$. What conjecture does Problem 19 suggest for the size of the coefficients of a function in $S(p)$?

21. The function

$$k(c, z) \equiv \frac{1}{2c}\left[\left(\frac{1+z}{1-z}\right)^c - 1\right] \tag{25}$$

is often called the *generalized Koebe function*. Show that if $c = 2$, this gives the original Koebe function. Prove that if $0 < c < 2$, then $k(c, z)$ is in S. If $0 < c < 2$, what is the image of E under $k(c, z)$? For more properties of this function see Campbell and Pfaltzgraff 1975 [2-B-221].

22. Find $f(E)$ for the function $f_2(z) = z/(1 - z^2)$.

23. Find the Maclaurin series for the function defined in Problems 22. Show that $f_2(z)$ is in $S^{(2)}$. What conjecture does this function suggest for the size of $|a_{2n+1}|$ for odd univalent functions?

24. The radius of univalence of a function $f(z)$ is denoted by R_S and is defined to be the radius of the largest open disk with center at the origin, in which $f(z)$ is univalent. Find R_S for $f(z) = z + z^2$.

25. Find R_S for $f(z) = z/(1+z)^3$.

26. Find R_S for $f(z) = z/(1+z)^4$.

★ **27.** Is it true that $R_S = 1/k$ if $f(z) = z/(1+z)^{k+1}$, when k is a positive integer? Note that it is not enough to prove that $f'(1/k) = 0$. This only proves that $R_S \leq 1/k$.

28. Find R_S for $f(z) = (a+bz)/(1-z)^2$ if a and b are positive.

29. Let a_n be the nth coefficient for the Maclaurin series for the functions given in Problem 28. Prove that if $a > 0$ and $b > 0$, then $|a_n| < n|a_1|$.

30. Find R_S for $f(z) = e^z$.

31. Let $h(z) = (1-az)/(1-z)$ where $a = Ae^{2i\alpha}$, $A > 0$, and $0 < \alpha < \pi$. Prove that $h(E)$ is a half plane H. Prove that: if $0 < A < 1$, then $w = 0$ is an exterior point of H; if $A > 1$, then $w = 0$ is an interior point of H; and, finally, if $A = 1$, then $w = 0$ is a boundary point of H. Find $h(e^{i\theta})$ in this case. What angle does this line make with the real axis?

★ **32.** Let $f(z)$ be the function obtained from $h(z)$ defined in Problem 31 when we square $h(z)$ and then normalize. Prove that

$$f(z) = \frac{z - cz^2}{(1-z)^2}, \tag{26}$$

and that $f(z)$ is univalent in E if and only if $|2c-1| \leq 1$. Prove that if $|2c-1| = 1$, then $c = e^{i\alpha}\cos\alpha$, where $0 < \alpha < \pi$. In this case $f(E)$ is the entire complex plane minus a straight line cut. Find the equation of the cut and the angle it makes with the real axis. Check your result by considering the case $\alpha = \pi/2$.

33. Do Problem 29 for the function defined in Problem 32.

★ **34.** Let $\lambda_1, \lambda_2, \ldots, \lambda_n, \ldots$ be a sequence of real numbers and let $\mu_1, \mu_2, \ldots, \mu_n, \ldots$ be a sequence of positive numbers. Set

$$F_1(z) = \left(\frac{1+z}{1-z} + i\lambda_1\right)^2, \qquad G_1(z) = (F_1(z) + \mu_1)^{1/2}. \tag{27}$$

For each $n > 1$, set

$$F_n(z) = (G_{n-1}(z) + i\lambda_n)^2, \qquad G_n(z) = (F_n(z) + \mu_n)^{1/2}. \tag{28}$$

Show that in defining $G_n(z)$ we can always select a branch so that

Re $G_n(z) > 0$ for z in E. Describe $F_n(E)$ geometrically. See Kakehashi 1959 [1-B-652], and 1967 [2-B-672].

★ **35.** Let D be a simply-connected domain in the w-plane that does not contain either of the two distinct points a and b. Prove that $\phi_1(w) \equiv w + \sqrt{(w-a)(w-b)}$ and $\phi_2(w) \equiv w - \sqrt{(w-a)(w-b)}$ are univalent in D. Thus if $f(z)$ is univalent in E and $f(E) \subset D$, then the composite functions $\phi_1(f(z))$ and $\phi_2(f(z))$ are univalent in E. Further, $f(z)$ is the sum of two different univalent functions:

$$f(z) = \frac{1}{2}\Big(\phi_1(f(z)) + \phi_2(f(z))\Big). \tag{29}$$

For interesting applications of these auxiliary functions, see Brickman 1970 [2-B-182].

★★ **36.** Let $f(z) = z + \Sigma_2^\infty a_k z^k$ be regular in E and suppose that there exists some N such that $|a_N| > 1/N$. Prove that there is some real ϕ such that the related function

$$g(z) = z + e^{i\phi} \sum_{k=2}^{\infty} a_n z^n \tag{30}$$

is not univalent in E.

Further, prove that there exists some real Φ such that the related function

$$G(z) = z + \sum_{k=2}^{N-1} a_k z^k + e^{i\Phi} a_N z^N + \sum_{k=N+1}^{\infty} a_k z^k \tag{31}$$

is not univalent in E. See Čin 1954 [1-B-211]. These two theorems have been generalized by Li 1964 [1-B-A-90].

CHAPTER 3

SOME AREA THEOREMS

Since the area of a domain is always positive, any expression for the area automatically supplies us with an inequality. It is a fortunate accident (?) that such inequalities can be used to prove some of the central theorems in the theory of univalent functions. Similar considerations about the length of an arc do not seem to be so fruitful. We devote this chapter to proving some of these area theorems. Applications of these theorems will be given in the chapters that follow.

Theorem 1. Fejér 1913 [3-B-342], and 1917 [1-B-291, p. 69]. Suppose that

$$f(z) = \sum_{n=0}^{\infty} a_n z^n \tag{1}$$

is regular in the closed disk $\overline{E_r}$: $|z| \leq r$. If $A(r, f) \equiv A(r)$ is the area of $f(\overline{E_r})$, then

(2) $$A(r) = \pi \sum_{n=1}^{\infty} n|a_n|^2 r^{2n}.$$

We note that in this theorem $f(z)$ is not assumed to be univalent. The area refers to the area of the Riemann surface $f(\overline{E_r})$, and if a plane domain D is covered k times by $f(\overline{E_r})$, then the area of D is counted k times in $A(r)$. The formula for $A(r)$ is "additive" over the terms. Thus if $f(z)$ is just the monomial $a_n z^n$, then $f(\overline{E_r})$ is an n-fold disk of radius $R = |a_n|r^n$ and from elementary geometry the area of this n-fold disk is $n\pi R^2 = n\pi|a_n|^2 r^{2n}$. Equation (2) is merely a sum of such terms. Thus the formula for $A(r)$ is easy to memorize.

Proof. If $w = u + iv$, and $z = x + iy$, then

(3) $$A(r) = \iint_{f(\bar{E}_r)} du\, dv = \iint_{\bar{E}_r} \frac{\partial(u, v)}{\partial(x, y)}\, dx\, dy,$$

by the usual rule for the transformation of a double integral. Under the conditions of the theorem, the partial derivatives are continuous and satisfy the Cauchy-Riemann equations. Hence,

(4) $$\frac{\partial(u, v)}{\partial(x, y)} = \begin{vmatrix} u_x & u_y \\ v_x & v_y \end{vmatrix} = \begin{vmatrix} u_x & -v_x \\ v_x & u_x \end{vmatrix} = u_x^2 + v_x^2 = |f'(z)|^2.$$

If we use (4) in (3) and change to polar coordinates ($z = x + iy = \varrho e^{i\theta}$), then

(5) $$A(r) = \iint_{|z| \le r} |f'(z)|^2\, dx\, dy = \int_{\rho=0}^{r} \int_{\theta=0}^{2\pi} |f'(\varrho e^{i\theta})|^2 \varrho\, d\theta\, d\varrho.$$

We observe that the term $|f'(z)|$ in equation (5) is a magnification factor and that the expression for $A(r)$ given by equation (5) is intuitively obvious from the fact that under the transformation $w = f(z)$ the "element of area $dx\, dy$" is transformed into the magnified "element of area $du\, dv = |f'(z)|^2\, dx\, dy$". For the integration on θ, we have

(6)
$$\int_0^{2\pi} |f'(\varrho e^{i\theta})|^2\, d\theta = \int_0^{2\pi} f'(\varrho e^{i\theta})\overline{f'(\varrho e^{i\theta})}\, d\theta$$
$$= \int_0^{2\pi} \left(\sum_{n=1}^{\infty} n a_n \varrho^{n-1} e^{(n-1)i\theta}\right)\left(\sum_{m=1}^{\infty} m\bar{a}_m \varrho^{m-1} e^{-(m-1)i\theta}\right) d\theta$$

$$= 2\pi \sum_{n=1}^{\infty} n^2 |a_n|^2 \varrho^{2n-2},$$

since (by the orthognality of the set $\{e^{in\theta}\}$) the terms with $m \neq n$ give 0 on integration. Using (6) in (5) we have

$$A(r) = \int_{\rho=0}^{r} 2\pi \sum_{n=1}^{\infty} n^2 |a_n|^2 \varrho^{2n-2} \varrho \, d\varrho = \pi \sum_{n=1}^{\infty} n |a_n|^2 r^{2n}. \quad \blacksquare$$

The same method gives

Theorem 2. PLE. Suppose that

$$f(z) = \sum_{n=-\infty}^{\infty} a_n z^n \tag{7}$$

is regular in the closed ring domain $D \equiv \bar{E}(r, R)$: $r \leq |z| \leq R$. Then the area of $f(D)$ is given by

$$A = \pi \sum_{n=-\infty}^{\infty} n |a_n|^2 (R^{2n} - r^{2n}). \tag{8}$$

Theorem 3. Let $f(z) = \Sigma_{-\infty}^{\infty} a_n z^n$ be regular and univalent on the circle C_r: $|z| = r$, and suppose that Γ, the image of C_r under $f(z)$, is described in a positive (counterclockwise) direction as θ runs from 0 to 2π. Then the area of the domain enclosed by Γ is given by

$$I = \pi \sum_{n=-\infty}^{\infty} n |a_n|^2 r^{2n}. \tag{9}$$

If Γ is described in a negative (clockwise) manner, then $I < 0$ and the area is $|I|$.

Proof. We introduce coordinates $w = u + iv$ in the w-plane, and we recall that if Γ is any simple closed positive curve in the w-plane parametrized by $u = u(t)$, $v = v(t)$ for $a \leq t \leq b$, then the area of the domain enclosed by Γ is

$$A = \frac{1}{2} \int_{\Gamma} (u \, dv - v \, du) = \frac{1}{2} \int_{a}^{b} \left(u(t) \frac{dv}{dt} - v(t) \frac{du}{dt} \right) dt. \tag{10}$$

In our case we select θ as the parameter ($z = re^{i\theta}$) and observe that

$$\text{(11)}\quad u(\theta) = \frac{1}{2}(f(z)+\overline{f(z)}) = \frac{1}{2}\sum_{n=-\infty}^{\infty}(a_n r^n e^{in\theta} + \overline{a_n} r^n e^{-in\theta}),$$

$$\text{(12)}\quad v(\theta) = \frac{1}{2i}(f(z)-\overline{f(z)}) = \frac{1}{2i}\sum_{n=-\infty}^{\infty}(a_n r^n e^{in\theta} - \overline{a_n} r^n e^{-in\theta}).$$

If we use (11) and (12) to compute $u'(\theta)$ and $v'(\theta)$ and substitute in equation (10), we will obtain a rather complicated looking expression which can be written as the sum of 8 integrals,

$$\text{(13)}\qquad A = \frac{1}{8}\sum_{k=1}^{8} I_k,$$

where each I_k has the general form

$$\text{(14)}\qquad I_k = \int_0^{2\pi}\sum_{n=-\infty}^{\infty}\sum_{m=-\infty}^{\infty} m b_n c_m r^{m+n} e^{\pm i(m\pm n)\theta}\, d\theta,$$

where b and c represent either a or $\bar{a}$. A close examination will show that in four of these integrals, where m and n occur in the form $i(m+n)$, the terms in the sum will cancel after integration and these integrals are 0. Each of the remaining four integrals yield $I_k = 2\pi\Sigma_{-\infty}^{\infty} n|a_n|^2 r^{2n}$. Using this in (13), we obtain (9).

If Γ is described in a negative manner, we obtain a positive description of the curve by replacing θ by $-\theta$. Then $A = -I$ and consequently $I < 0$. ■

Corollary 1. Let

$$\text{(15)}\qquad f(z) = cz + c_0 + \sum_{n=1}^{\infty}\frac{c_n}{z^n}, \qquad c \neq 0,$$

be regular and univalent in $\bar{E}_r^x$: $r \leq |z| < \infty$. Then the area of the set ommited[1] by $f(\bar{E}_r^x)$ is

$$\text{(16)}\qquad A = \pi\left(|c|^2 r^2 - \sum_{n=1}^{\infty}\frac{n|c_n|^2}{r^{2n}}\right).$$

The situation is indicated in Figure 1. Here $\bar{E}_r^x$ and its image are

[1] The term "ommited set" means the complement of $f(\bar{E}_r^x)$.

shaded. The series in (16) gives the area of the unshaded region in the w-plane.

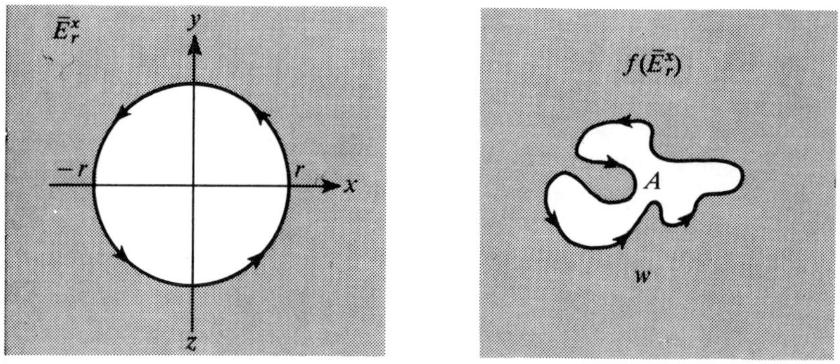

FIGURE 1.

Proof. If $f(z)$ is univalent, then for large $\varrho > r$ the image of the circle C_ρ under $f(z)$ is close to the circle $w = c\varrho e^{i\theta}$. Hence $\Gamma_\rho \equiv f(C_\rho)$ is described in a positive direction. Since $f(z)$ is continuous and is univalent, this positive orientation must be preserved for all $\varrho \geq r$. Then Γ_r is a positive curve. We apply Theorem 3 with $a_{-n} = c_n$ $(n = 1, 2, \ldots)$ and $a_1 = c$. Then equation (9) yields equation (16). ■

We observe that negative terms occur in equation (9). Equation (16) makes these negative terms visible (by a change in notation). Further, under the conditions of Theorem 3 we have the inequality

$$0 \leq |c|^2 r^2 - \sum_{n=1}^{\infty} \frac{n|c_n|^2}{r^{2n}}. \tag{17}$$

By a suitable argument we can remove the equal sign from (17) but we have no need for this sharpened form of the inequality.

Theorem 4. Gronwall 1914 [1-B-452, p. 75]. The Area Theorem. If

$$f(z) = z + \sum_{n=0}^{\infty} \frac{c_n}{z^n} \tag{18}$$

is regular and univalent in E^x: $|z| > 1$, then

$$\sum_{n=1}^{\infty} n|c_n|^2 \leq 1. \tag{19}$$

Further, the equality sign in (19) holds if and only if $f(E^x)$ covers the complex plane except for a set of zero area (the complement of $f(E^x)$ contains no open set).

This theorem has the special title *the Area Theorem* because it is so important in the theory of univalent functions. To emphasize this importance many writers use the German equivalent *Flächensatz*.

Proof. The conditions of Corollary 1 are satisfied for each $r > 1$ and hence we have (17) with $c = 1$. We divide (17) by r^2, and let $r \to 1^+$. Then the infinite series is an increasing function of r bounded above by 1. Hence the series converges at $r = 1$ and satisfies the inequality (19). We leave to the reader the discussion of the case when the equality sign holds. ■

PROBLEMS

1. Prove Theorem 2. Is equation (8) "additive" over the terms?
2. Complete the computations required in the proof of Theorem 3.
3. Suppose that $\epsilon > 0$. Let Γ be the image of $z = e^{i\theta}$ $(0 \leq \theta \leq 2\pi)$ under

$$f(z) \equiv \frac{1+\epsilon}{z^2} + \frac{1}{z} + z + (1-\epsilon)z^2.$$

Prove that either Γ has a self-intersection point or Γ has a negative direction as the parameter θ increases.

4. Generalize the result stated in Problem 3.
5. Prove that for each $R > 1$, the image of C_R under $f(z) = z + 1/z$ is an ellipse. If a and b are the semi-major and semi-minor axes, find a and b. What is the limit of this ellipse as $R \to 1$?
6. Use Theorem 3 to compute the area enclosed by the ellipse described in Problem 5.
7. Recall that HZ is the class of all function $f(z) = z + \Sigma_{n=2}^{\infty} a_n z^n$ for which

$$\sum_{n=2}^{\infty} n|a_n| \le 1. \tag{20}$$

In Problem 2.7 we saw that each function in HZ is univalent in E. For each fixed r let

$$\begin{cases} M(r,f) \equiv \max |f(z)|, & \text{for } |z| = r, \\ m(r,f) \equiv \min |f(z)|, & \text{for } |z| = r. \end{cases} \tag{21}$$

Prove that if $f(z)$ is in HZ and $0 \le r \le 1$, then $M(r,f) \le r + r^2/2$ and $m(r,f) \ge r - r^2/2$. Further, these bounds can be obtained by some function in the class.

8. Prove that if $f(z)$ is in HZ, then $f(E)$ always contains the disk $|w| < 1/2$ and is always contained in the disk $|w| < 3/2$.
9. Under the conditions of Problem 8, find the maximum of A, where A is the area of $f(E)$.
10. Prove that if $f(z)$ is in S, then $A \ge \pi$, where A is the area of $f(E)$. Show that $A = \pi$ if and only if $f(z) = z$.
11. The computation for Theorem 3 may be shortened if we use polar coordinates $z = re^{i\theta}$ and $w = Re^{i\Phi}$ and one of the corresponding Cauchy-Riemann equations, equation (1.33),

$$\frac{\partial \Phi}{\partial \theta} = \frac{r}{R}\frac{\partial R}{\partial r}.$$

Fill in the missing steps in the following outline.

$$A = \frac{1}{2}\int R^2\, d\Phi = \frac{1}{2}\int_0^{2\pi} R^2 \frac{\partial \Phi}{\partial \theta}\, d\theta = \frac{1}{2}\int_0^{2\pi} R^2 \frac{r}{R}\frac{\partial R}{\partial r}\, d\theta$$

$$= \frac{r}{4}\int_0^{2\pi} \frac{\partial}{\partial r} |f(re^{i\theta})|^2\, d\theta$$

$$= \frac{r}{4}\int_0^{2\pi} \sum_{n=-\infty}^{\infty}\sum_{m=-\infty}^{\infty} (n+m)a_n\bar{a}_m r^{n+m-1} e^{i(n-m)\theta}\, d\theta$$

$$= \pi \sum_{n=-\infty}^{\infty} n|a_n|^2 r^{2n}.$$

12. Prove that if $f(z) = \sum_{n=0}^{\infty} a_n z^n$ is regular in E and $r < 1$, then

$$\frac{1}{2\pi}\int_0^{2\pi} |f(re^{i\theta})|^2\, d\theta = \sum_{n=0}^{\infty} |a_n|^2 r^{2n}.$$

This identity is sometimes called Gutzmer's Theorem [1]. See Tammi 1957 [1-B-1382].

13. Prove that if $|f(z)| \leq M$ in E, then

$$\sum_{n=0}^{\infty} |a_n|^2 \leq M^2$$

and hence the series on the left side converges. Show that this generalizes the Cauchy inequality.

14. Let Γ_r be the image of the circle $|z| = r$, under $f(z)$, and $L(r, f)$ be the length of Γ_r. If $f(z)$ is regular in E and $r < 1$, prove analytically that

$$L(r, f) = \int_0^{2\pi} r|f'(z)|\, d\theta, \qquad \text{where } z = re^{i\theta}.$$

Observe that the formula is geometrically obvious because $|f'(z)|$ is the magnification factor at each point of the circle $|z| = r$.

15. Under the conditions of Problem 14 prove that

$$L(r, f) \leq 2\pi r \sqrt{\sum_{n=1}^{\infty} n^2|a_n|^2 r^{2n-2}}\,.$$

16. Let $L(\alpha, r, f)$ be the length of the image of the radial segment $z = te^{i\alpha}$, $0 \leq t \leq r$ under $f(z)$. Find a formula for $L(\alpha, r, f)$.

17. In some of the proofs in this chapter it was necessary to interchange the order of integration and summation of an infinite series. Locate the places where the interchange occurred, and in each case justify that step.

CHAPTER 4

ELEMENTARY BOUNDS FOR THE COEFFICIENTS

1. THE SECOND COEFFICIENT IN THE SET *S*

This sharp bound is given in

Theorem 1. Bieberbach 1916 [1-B-117]. If[1]

$$f(z) = z + \sum_{n=2}^{\infty} a_n z^n \tag{1}$$

is in S, then

$$|a_2| \leq 2 \tag{2}$$

[1] To avoid repetition of the series (1) we agree that whenever we write $f(z)$ is in S, we assume that $f(z)$ has the series expansion (1) unless some other symbols are explicitly stated.

Further, this inequality is sharp. Equality occurs in (2) if and only if $f(z)$ is a rotation of the Koebe function:

$$f(z) = e^{-i\alpha}k(e^{i\alpha}z) = \frac{z}{(1-e^{i\alpha}z)^2}. \tag{3}$$

An inequality is *sharp* (French *exact*, German *genauer*, Russian *tochni*) if, under the conditions given, it is impossible to improve the inequality (decrease an upper bound, or increase a lower bound) because there is an admissible function for which the equal sign holds. In Theorem 1 the inequality (2) is sharp, because the Koebe function is in S and for this function $a_2 = 2$. Although this is the first time we have used the term *sharp*, we have met such inequalities already (Theorem 3.4 and Problems 2.6, 2.7, 3.7, 3.8, 3.9, and 3.10).

A function for which equality occurs is called an *extremal* function. Thus, in Theorem 1, the functions given by (3) are extremal functions.

Proof. Let $g(z) = \sqrt{f(z^2)}$. Then by Theorem 2.2

$$g(z) = z + \frac{1}{2}a_2 z^3 + \left(\frac{1}{2}a_3 - \frac{1}{8}a_2^2\right)z^5 + \cdots \tag{4}$$

is in S. By Theorem 2.6 the function

$$\phi(\zeta) = \frac{1}{g(1/\zeta)} = \zeta - \frac{1}{2}a_2\frac{1}{\zeta} + c_3\frac{1}{\zeta^3} + c_5\frac{1}{\zeta^5} + \cdots \tag{5}$$

is in Σ_0. By the Area Theorem (Theorem 3.4)

$$\sum_{n=1}^{\infty} n|c_n|^2 = \left|-\frac{a_2}{2}\right|^2 + 3|c_3|^2 + 5|c_5|^2 + \cdots \le 1. \tag{6}$$

Hence $|-a_2/2|^2 \le 1$, and this gives $|a_2| \le 2$. If $a_2 = 2e^{i\alpha}$, then in the inequality (6) we have $c_n = 0$ for all $n \ge 2$. Hence (5) becomes $\phi(\zeta) = \zeta - e^{i\alpha}/\zeta$. Thus

$$g(z) = \frac{1}{\phi(1/z)} = \frac{1}{\dfrac{1}{z} - e^{i\alpha}z} = \frac{z}{1-e^{i\alpha}z^2}$$

Since $f(z^2) = g^2(z) = z^2/(1-e^{i\alpha}z^2)^2$, we find that $f(z) = z/(1-e^{i\alpha}z)^2$. ∎

2. SOME OTHER COEFFICIENT BOUNDS

Suppose that in the proof of Theorem 1 we omit the intermediate step of considering $g(z) = \sqrt{f(z^2)}$ and replace equation (5) by $\phi(\zeta) = 1/f(1/\zeta)$. Then we will obtain

Theorem 2. PLE. Bieberbach 1916 [1-B-117]. If $f(z)$ is in S, then

$$|a_2^2 - a_3| \leq 1. \tag{7}$$

This inequality is sharp. Equality occurs in (7) if and only if $f(z)$ is a rotation of the function

$$\tilde{k}_\phi(z) = \frac{z}{1 - bz + z^2}, \qquad -2 \leq b \leq 2, \qquad b = 2\cos\phi. \tag{8}$$

Inequality (7) together with $|a_2| \leq 2$ gives immediately the crude bound $|a_3| \leq 5$ for any $f(z)$ in S. Of course, the central conjecture is that $|a_3| \leq 3$, so this inequality is not very exciting. The novice might hope that by a preliminary rotation he or she can introduce an arbitrary parameter in the expression $|a_2^2 - a_3|$ and thus prove that $|a_3| \leq 3$ for $f(z)$ in S. Unfortunately $|a_2^2 - a_3|$ is invariant under the rotation $e^{-i\alpha}f(e^{i\alpha}z)$.

Corollary 1. PLE. If $f(z)$ is in S and $a_2 = 0$, then

$$|a_3| \leq 1, \tag{9}$$

and this inequality is sharp. Equality occurs if and only if

$$f(z) = \frac{z}{1 - e^{i\alpha}z^2}, \qquad \alpha \text{ real.} \tag{10}$$

Since $f(z) = z/(1 - e^{i\alpha}z^2)$ is an odd function, this corollary also solves the problem of maximizing $|a_3|$ in the class $S^{(2)}$ of odd univalent functions.

Theorem 3. Suppose that $k > 1$ is an integer and

$$g(z) \equiv z + b_{k+1}z^{k+1} + b_{2k+1}z^{2k+1} + \cdots = z + \sum_{n=1}^{\infty} b_{nk+1}z^{nk+1} \tag{11}$$

is in $S^{(k)}$. Then

(12) $$|b_{k+1}| \leq \frac{2}{k},$$

and this inequality is sharp. Equality occurs if and only if $f(z)$ is a rotation of

(13) $$g(z) = \frac{z}{(1-z^k)^{2/k}}.$$

Proof. Since $g(z)$ is in $S^{(k)}$, there is an $f(z)$ in S such that

$$\begin{aligned} g(z) &= [f(z^k)]^{1/k} = [z^k + a_2 z^{2k} + a_3 z^{3k} + \cdots]^{1/k} \\ &= z[1 + a_2 z^k + a_3 z^{2k} + \cdots]^{1/k} \\ &= z + \frac{1}{k} a_2 z^{k+1} + \left(\frac{1}{k} a_3 - \frac{1}{2}\,\frac{k-1}{k^2}\, a_2^2\right) z^{2k+1} + \cdots. \end{aligned}$$

Hence $b_{k+1} = a_2/k$. Since $|a_2| \leq 2$ for $f(z)$ in S, it is obvious that $|b_{k+1}| \leq 2/k$. Equality occurs in Theorem 1 if and only if $f(z)$ is a rotation of the Koebe function. Hence equality occurs in (12) if and only if $g(z) = [f(z^k)]^{1/k}$ when $f(z)$ is the Koebe function. This gives equation (13). ■

For a generalization of this theorem see Problem 17.11.

3. BOUNDED UNIVALENT FUNCTIONS

We can also obtain the sharp bound for the second coefficient in the set of bounded univalent functions.

Definition 1. Let $SB_1(M)$ denote the set[1] of all functions in S for which

(14) $$|f(z)| < M$$

for all z in E.

From Schwarz's Lemma (see Theorem 7.8 or any text on Function Theory) it is easy to see that if $M < 1$, the set $SB_1(M)$ is empty and if

[1] For an explanation of this rather complicated looking notation for the set, see Chapter 12.1.

$M = 1$, then $SB_1(M)$ contains only one function $f(z) \equiv z$. Hence we may assume that $M > 1$. Before we find the sharp bound for $|a_2|$ in the set $SB_1(M)$, we construct the extremal function.

Let $z = K(w)$ be the inverse function for the Koebe function $k(z)$. If $0 < t < 1$, then the function $tk(z)$ maps E onto the entire complex plane except for a slit along the negative real axis from $-\infty$ to $-t/4$. Now $K(w)$ is regular in this domain, and carries it back into the unit disk. In fact the image will cover E except for the slit on the negative real axis from -1 to $K(-t/4)$. Thus, if we normalize this composite function, we obtain $F(z) \equiv K(tk(z))/t$, and $F(z)$ maps E as indicated in Figure 1.

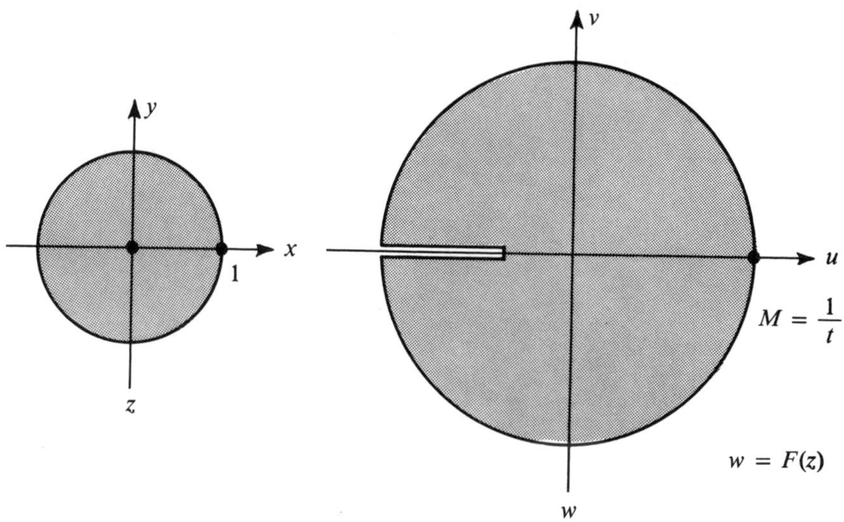

FIGURE 1

It is easy to see that

$$K(w) = \frac{2w + 1 - \sqrt{1+4w}}{2w}, \tag{15}$$

where the radical is 1 when $w = 0$. Expanding the radical we find that

$$\begin{aligned} K(w) &= w + \sum_{n=2}^{\infty} \frac{(-1)^{n+1} 1 \cdot 3 \cdot 5 \cdots (2n-1) 2^n}{(n+1)!} w^n \\ &= \sum_{n=1}^{\infty} \frac{(-1)^{n+1}(2n)!}{(n+1)!\,n!} w^n. \end{aligned} \tag{16}$$

We observe in passing that, although the coefficient conjecture $|a_n| \leq n$ for $f(z)$ in S is still open, the sharp upper bound for inverse functions is known. Indeed, in Chapter 16.7 we will give Baernstein's proof of Loewner's Theorem that if

$$z = w + \sum_{n=2}^{\infty} b_n w^n$$

is the inverse of a function $f(z)$ in S, then for $n = 2, 3, 4, \ldots$

$$\text{(17)} \qquad |b_n| \leq \frac{1}{n}\binom{2n}{n+1} = \frac{(2n)!}{(n+1)!n!},$$

with equality if and only if $f(z)$ is a rotation of the Koebe function.

We continue our computation of the function that maps E as indicated in Figure 1. From $k(z) = z/(1-z)^2$ and equation (15) we find that

$$\text{(18)} \qquad F(z) = \frac{1}{t}K(tk(z))$$
$$= \frac{z}{(1-z)^2} + \sum_{n=2}^{\infty} \frac{(-1)^{n+1}(2n)!}{(n+1)!n!} \frac{t^{n-1}z^n}{(1-z)^{2n}}.$$

Since $K(tk(z))$ maps E onto E with a slit, it follows that $|F(z)| < 1/t$ for z in E. Hence $M = 1/t$. Using this in (18) we have

(19)

$$F(z) = M\frac{M(1-z)^2 + 2z - (1-z)\sqrt{M^2(1-z)^2 + 4Mz}}{2z} \equiv k(z, M).$$

This special function $k(z, M)$ is sometimes called the *bounded Koebe function* (see Figure 1). Equation (18) gives

$$\text{(20)} \qquad k(z, M) = z + 2\left(1 - \frac{1}{M}\right)z^2 + \left(1 - \frac{1}{M}\right)\left(3 - \frac{5}{M}\right)z^3$$
$$+ \left(1 - \frac{1}{M}\right)\left(4 - \frac{16}{M} + \frac{14}{M^2}\right)z^4 + \cdots.$$

Theorem 4. Pick 1917 [1-B-1049]. If $f(z)$ is in $SB_1(M)$, then

$$\text{(21)} \qquad |a_2| \leq 2(1 - 1/M).$$

Equality occurs if and only if $f(z)$ is a rotation of the function $k(z, M)$ defined by equations (18) and (19) with $t = 1/M$.

The proof of the theorem is shorter than the construction of the example function.

Proof. If $|f(z)| < M$, then $k(f(z)/M)$ is regular and univalent in E. To normalize we multiply by M. Hence

$$\text{(22)}\quad g(z) = Mk\left(\frac{f(z)}{M}\right) = M\left[\frac{f(z)}{M} + 2\frac{f^2(z)}{M^2} + 3\frac{f^3(z)}{M^3} + \cdots\right]$$

$$= z + a_2z^2 + a_3z^3 + \cdots + \frac{2}{M}[z^2 + 2a_2z^3 + \cdots]$$

$$+ \frac{3}{M^2}[z^3 + 3a_2z^4 + \cdots] + \cdots$$

$$= z + \left(a_2 + \frac{2}{M}\right)z^2 + \left(a_3 + \frac{4a_2}{M} + \frac{3}{M^2}\right)z^3 + \cdots$$

is in S. Therefore $|a_2 + 2/M| \leq 2$ with equality if and only if $g(z)$ is a rotation of $k(z)$. In this situation, we can make a preliminary rotation of $f(z)$ so that $a_2 \geq 0$. Consequently we can write that $a_2 + 2/M \leq 2$ and this gives (21). ■

What about $|a_3|$ for $f(z)$ in $SB_1(M)$? Here the superficial guess that the same function supplies the maximum for $|a_3|$ is *trivially wrong*. For a glance at equation (20) shows that the third coefficient is zero when $M = 5/3$, and 0 cannot possibly be the upper bound for $|a_3|$ in a class of functions that contains $z + z^3/3$. The difficult problem of finding the sharp upper bound for $|a_3|$ in the class $SB_1(M)$ was solved independently by Schaeffer and Spencer 1945 [1-B-1234, p. 114] and Tammi 1953 [1-B-1377, p. 9] and [1-B-1378]. Schaeffer and Spencer only state the result. Using Loewner's method, Tammi proved

Theorem 5. PWO. If $f(z)$ is in $SB_1(M)$, then

$$\text{(23)}\qquad |a_3| \leq \begin{cases} 1 - \dfrac{1}{M^2}, & \text{if } 1 \leq M \leq e, \\ 2\left(\sigma - \dfrac{1}{M}\right)^2 + 1 - \dfrac{1}{M^2}, & \text{if } e \leq M, \end{cases}$$

where σ is the real root of $Mx \ln x = -1$.

For other proofs of Theorem 5, see Janowski 1955 [1-B-586] and V. Singh 1959 [1-B-1309].

4. UNIVALENT FUNCTIONS THAT HAVE A POLE

Here we consider the class of functions $S(p)$ introduced in Problems 2.18, 2.19, and 2.20. This is the class of normalized functions that are univalent in E and have a pole at z_0 where $0 < |z_0| = p < 1$.

Theorem 6. Komatu 1945 [1-B-709]. If $f(z)$ is in $S(p)$, then

$$|a_2| \leq \frac{1+p^2}{p}. \tag{24}$$

Equality occurs if and only if $f(z)$ is a rotation of the function

$$k_p(z) = \frac{z}{1-bz+z^2}, \qquad b = p + \frac{1}{p} > 2. \tag{25}$$

Earlier Fenchel 1931 [1-B-307] proved that if $|a_2| > 2$, then $p \leq 2/(|a_2| + \sqrt{|a_2^2| - 4})$, and this result will give (24), see Theorem 17.35.

Proof. By a rotation we may place the pole at $-p < 0$. The function $K(tk(z))$, used in the proof of Theorem 4, has the expansion

$$\begin{aligned} K(tk(z)) &= \sum_{n=1}^{\infty} \frac{(-1)^{n+1}(2n)!}{(n+1)!n!} \frac{t^n z^n}{(1-z)^{2n}} \\ &= tz + 2t(1-t)z^2 + t(1-t)(3-5t)z^3 + \cdots, \end{aligned} \tag{26}$$

and maps E onto E with a slit from -1 to $K(-t/4)$. We determine t so that the slit ends at $-p$. This gives $t = -4k(-p) = 4p/(1+p)^2$. On the other hand, $f(K(tk(z)))$ is regular and univalent in E. If we normalize we find that

$$\begin{aligned} \frac{1}{t} f(K(tk(z))) &= \frac{1}{t}[tz + 2t(1-t)z^2 + \cdots + a_2(tz + \cdots)^2 + \cdots \\ &= z + (2(1-t) + a_2 t)z^2 + \cdots. \end{aligned}$$

Since $0 < t = 4p/(1+p)^2 < 1$ we can write

$$t|a_2| \le 2 + 2(1-t) = 4 - 2t$$
$$|a_2| \le \frac{4}{t} - 2 = \frac{1+2p+p^2}{p} - 2 = \frac{1+p^2}{p}. \quad \blacksquare$$

Using similar considerations Jenkins 1962 [1-B-629] has proved

Theorem 7. PWO. If $f(z)$ is in $S(p)$, then

$$|a_n| \le \frac{1+p^2+p^4+\cdots+p^{2n-2}}{p^{n-1}} \tag{27}$$

is true for $n = 2, 3, 4, 5,$ and 6.

It is clear that if $p \to 1$, the inequality (27) yields the bound $|a_n| \le n$ for the class S. Conversely Jenkins proved that if we know $|a_n| \le n$ for $n = 2, 3, 4, \ldots, N$ for the class S, then the inequality (27) is true for the class $S(p)$ for the same set of indices. Theorem 7 then follows from the fact that in the class S, the bound $|a_n| \le n$ has been proved for $n = 2, 3, 4, 5,$ and 6.

★ 5. DOES A PROBLEM HAVE A SOLUTION?

Let us return for a moment to our state of ignorance before we proved Theorem 1. We now ask for the maximum of $|a_2|$ for the set of functions in S. Logically there are several possibilities: (a) There is some maximum A_2, and there is a function in S for which $|a_2| = A_2$; (b) there is a least upper bound A_2 for the set of all $|a_2|$, but there is no $f(z)$ in S for which $|a_2| = A_2$; (c) the set of all $|a_2|$ for $f(z)$ in S is unbounded.

If case (c) occurs, we say that the problem of finding max $|a_2|$ for $f(z)$ in S *has no solution*. If case (a) or (b) occurs, then the problem has a solution, but it is desirable to know which of these two cases represents the truth. Once we have proved Theorem 1, all of the above questions are answered. The problem has a solution, $A_2 = 2$, and we know all of the extremal functions.

But if we turn to a more difficult problem, then all three of the cases (a), (b), and (c) may seem equally likely at the outset. What we need is a general theory that will help us to settle the matter. Here we outline the basic items for a theory that will tell us when a solution exists, and if it does exist whether case (a) or (b) holds. In this section we omit all proofs.

These can be found (for example) in the superb text by Nehari [1-B-978, pp. 137–147]. We can bypass this material because for most of the problems posed in this text it is intuitively clear that a solution exists. Further proving the required theorem (as we did in Theorem 1) rigorously removes all doubt.

Definiton 2. A functional defined on a family of functions $\mathcal{F}$ is a function whose domain is $\mathcal{F}$ and whose range is some subset of the set of all complex numbers. The functional is usually denoted by J and its value at f is $J[f]$. If the range of J is a subset of the real numbers, then J is called a real-valued functional.

Let $\mathcal{F} = S$ and let z_0 be a fixed point in E. Then $J_1[f] = f(z_0)$ is a functional. Other examples, using the same family, are:

$$J_2[f] = a_7, \qquad J_3[f] = |a_7 - a_3a_4 + a_2^2|, \tag{28}$$

$$J_4[f] = \operatorname{Re} f(z_0), \qquad J_5[f] = \arg f'(z_0), \tag{29}$$

$$J_6[f] = \frac{1}{2\pi}\int_0^{2\pi} |f''(re^{i\theta})|\, d\theta, \qquad r \text{ fixed in } (0, 1). \tag{30}$$

A functional J is *linear* if for every complex λ, μ and every f, g in $\mathcal{F}$

$$J[\lambda f + \mu g] = \lambda J[f] + \mu J[g]. \tag{31}$$

For example, J_1 and J_2 defined above are both linear, but the other functionals are not.

A real-valued functional is *convex* if for every λ in $[0, 1]$ and every f, g in $\mathcal{F}$

$$J[\lambda f + (1-\lambda)g] \le \lambda J[f] + (1-\lambda)J[g]. \tag{32}$$

For example, J_4 and J_6 are convex.

A functional J is said to be a *continuous functional* on $\mathcal{F}$ if for every sequence $\{f_n\}$ of functions in $\mathcal{F}$ that converges to an f in $\mathcal{F}$, we have

$$\lim_{n\to\infty} J[f_n] = J[f]. \tag{33}$$

We now look more closely at the family $\mathcal{F}$. Henceforth a family of functions will consist of some subset of all functions *regular* in some fixed domain D.

A family $\mathscr{F}$ of functions, each regular in a fixed domain D, is said to be *uniformly bounded* in D if there is an M such that for every f in $\mathscr{F}$ and every z in D

$$|f(z)| \leq M. \tag{34}$$

The family $\mathscr{F}$ is said to be *locally uniformly bounded* in D if each point z_0 in D has a neighborhood $\mathscr{N}$ in which $\mathscr{F}$ is uniformly bounded.

For example, if D is the unit disk and $\mathscr{F}$ is the set of functions $f_\alpha(z) = 1/(z - e^{i\alpha})$, α real, then $\mathscr{F}$ is not uniformly bounded, but $\mathscr{F}$ is locally uniformly bounded.

Theorem 8. PLE. $\mathscr{F}$ is locally uniformly bounded in D if and only if $\mathscr{F}$ is uniformly bounded for each compact subset of D.

Theorem 9. PLE. If $\mathscr{F}$ is a locally uniformly bounded family of functions each regular in D, then the family $\mathscr{F}^{(n)}$ of functions $f^{(n)}(z)$, where $f(z)$ is in $\mathscr{F}$, is locally uniformly bounded.

Definition 3. A family of functions $\mathscr{F}$ is said to be *equicontinuous* in D if for each $\epsilon > 0$ and each compact subset X of D, there is a $\delta > 0$ such that

$$|f(z_1) - f(z_2)| < \epsilon \tag{35}$$

for all $f(z)$ in $\mathscr{F}$ and for all pairs z_1, z_2 in X for which $|z_1 - z_2| < \delta$.

Theorem 10. PLE. If $\mathscr{F}$ is a locally uniformly bounded family of functions each regular in D, then $\mathscr{F}$ is equicontinuous in D.

The converse is false. The family $\mathscr{F}$ of functions $f_n(z) = z + n$ is equicontinuous in E but is not locally uniformly bounded. However, we have

Theorem 11. PLE. If $\mathscr{F}$ is a family of functions equicontinuous in D and if the set $\{f(z_0)\}$ is bounded for one fixed point z_0 in D, then $\mathscr{F}$ is locally uniformly bounded in D.

The really central concept is

Definition 4. A family $\mathscr{F}$ of functions each regular in D is said to be *normal* if every infinite sequence $\{f_n(z)\}$ of functions from $\mathscr{F}$ has a subsequence that converges uniformly on every compact subset of D.

Thus if b is fixed in E the family $f_\alpha(z) = e^{i\alpha}(z - b)/(1 - \bar{b}z)$, with α real, is a normal family. To help in finding families we need

Theorem 12. PLE. Suppose that $\{f_n(z)\}$ is a sequence of regular functions from a family of functions that is locally uniformly bounded in D. If the sequence converges at each point of a set X that is dense in D, then the sequence converges uniformly on every compact set contained in D.

To find normal families we use

Theorem 13. PWO. Montel's Theorem. A family of regular functions $\mathcal{F}$ that is locally uniformly bounded in D is a normal family in D.

In Chapter 6 we will prove a theorem (Theorem 6.8) that implies that S is a locally uniformly bounded family in E. Hence S is a normal family. Consequently any sequence of functions from S has a subsequence that is uniformly convergent in each disk $|z| \leq r < 1$.

The sequence $f_n(z) \equiv z/n$ of univalent functions certainly converges uniformly in E, but the limit function $f(z) \equiv 0$ is not univalent. Such an example cannot arise in the set S, because of the normalization $f'(0) = 1$. Thus we have

Theorem 14. PLE. If $\{f_n(z)\}$ is a sequence of functions from S that converges in any disk $|z| \leq r$, where $r > 0$, then the limit function is also in S.

Definition 5. A family of functions $\mathcal{F}$ is said to be *compact* if every convergent sequence of functions from $\mathcal{F}$ converges to a function that is also in $\mathcal{F}$.

For example, S is a compact family by Theorem 14.

We now have the machinery to treat the general question raised at the beginning of this section. Given a family of functions $\mathcal{F}$ and a real-valued functional J defined on $\mathcal{F}$, find the maximum value of J for functions in $\mathcal{F}$. (In Theorem 1, $\mathcal{F} = S$ and $J[f] = |a_2|$.)

First we must prove that the functional J is bounded above in $\mathscr{F}$. This is usually easy, but one must be careful, because it can happen that J is not bounded above, and then the problem has no solution.

If J is bounded above, let J_0 be the least upper bound of the set $J[f]$ for all f in $\mathscr{F}$. Then by the definition of J_0 there is a sequence of functions $\{f_n(z)\}$ from $\mathscr{F}$ for which

$$J_0 - 1/n < J[f_n] \leq J_0. \tag{36}$$

If the family $\mathscr{F}$ is normal in D, then we can select a convergent subsequence from the sequence $\{f_n(z)\}$, and if $\mathscr{F}$ is compact, this subsequence will converge to an $f(z)$ in $\mathscr{F}$. Finally, if the functional J is continuous, then $J[f] = J_0$. Thus our problem, find max $J[f]$ in the family $\mathscr{F}$, will always have at least one solution if we know: (a) J is a bounded real-valued continuous functional, (b) $\mathscr{F}$ is a normal family, and (c) $\mathscr{F}$ is a compact family (all on a fixed domain D).

As we have mentioned earlier, we do not need this elaborate theory for most of the work in this book. However, if we wish to use variational methods, then this theory becomes very important. In the *variational method* we start with an $f(z)$ in $\mathscr{F}$ and find a formula of the form

$$f^*(z) = f(z) + \epsilon Q(z) + o(\epsilon) \tag{37}$$

for a neighboring function $f^*(z)$ in $\mathscr{F}$. Suppose that we have a formula of the form (37). Then we say "let $f(z)$ be an extremal function that maximizes $J[f]$ and consider $J[f^*] \leq J[f]$." But *we cannot make this first move* unless we know that such an f exists. Hence the theory just outlined (or some equivalent theory) is absolutely necessary for the application of a variational formula.

PROBLEMS

1. Suppose that b satisfies the inequality $-2 < b < 2$. Find $f(E)$ for $f(z) = \widetilde{k}_\phi(z) \equiv z/(1 - bz + z^2)$. Find $\lim f(z)$ as $b \to \pm 2$. Hint: Consider $1/f(z)$ and use Problem 3.5.
2. Where are the poles for the function defined in Problem 1?
3. Find a_2 and a_3 for $f(z) = z/(1 - bz + z^2)$ and show that $a_2^2 - a_3 = 1$, for arbitrary b, real or complex.

4. Show that for any Maclaurin series the expression $|a_2^2 - a_3|$ is invariant under the rotation defined in Theorem 2.2 (Equation 2.7).
5. Prove Theorem 2.
6. Prove Corollary 1.
7. Let $g(w) = w + \Sigma_2^\infty b_n w^n$ be the inverse of a function in S. Prove that

$$b_2 = -a_2, \qquad b_3 = 2a_2^2 - a_3, \qquad b_4 = 5a_2a_3 - 5a_2^3 - a_4.$$

   Formulas for computing all of the b_n in terms of the a_n will be given in the next chapter (Theorems 5.6 and 5.7).
8. Prove that if $g(w)$ is the inverse of a function in S, then $|b_2| \le 2$. Suppose that we know that $|a_3| \le 3$ and $|a_4| \le 4$ for $f(z)$ in S. Prove that then $|b_3| \le 5$ and $|b_4| \le 14$ and that these two bounds are sharp. See thc inequality (17).
9. Referring to the bounds for $|a_3|$ given in Theorem 5 prove that both parts give the same bound when $M = e$. Find the limit of the bound as $M \to \infty$.
10. Prove that if $f(z)$ is in $S(p)$, then $|a_2^2 - a_3| \le 1$. As we know, the coefficients a_2^2 and a_3 may be arbitrarily large if p is close enough to 0, but a_2^2 and a_3 must be large together.
11. Find the sharp upper bound for $|b_2^2 - b_3|$ if $g(w)$ is the inverse of a function in S.
12. ★ Prove that if $g(z)$ is in $S^{(k)}$ and has the Maclaurin series (11), then

$$\left| b_{k+1}^2 - \frac{2}{k+1}\, b_{2k+1} \right| \le \frac{2}{k(k+1)},$$

   and this inequality is sharp.
13. ★ Prove that if $f(z)$ is in $SB_1(M)$, then

$$|a_2^2 - a_3| \le 1 - \frac{1}{M^2},$$

   and this inequality is sharp.
14. Check that equations (19) and (20) are correct.
15. Prove that in equation (20) the coefficient a_n is a polynomial in $1/M$ of degree $n-1$. Explain why $1 - 1/M$ is always a factor of this polynomial.
16. Let $\{f_n(z)\}$ be a sequence of functions each regular in a domain D, and suppose that $f_n(z) \to f(z)$ uniformly in D. Prove that either $f(z) \equiv 0$, or if $f(z)$ has a kth order zero for some z_0 in D, then there

is an $\epsilon_0 > 0$ such that for each ϵ in $(0, \epsilon_0)$ there is an $N(\epsilon)$ such that in $|z - z_0| < \epsilon$ every function $f_n(z)$ with $n > N(\epsilon)$ has k zeros when counted in accordance with their multiplicities. This is called Hurwitz's Theorem. See Hille [1-B-537, vol. II p. 205].

17. State and prove a converse for the theorem in Problem 16.

★ **18.** Suppose $f(z)$ is in S and has the power series (1). Let $R_S(N)$ be the radius of univalence of the polynomial $P_N(z) \equiv z + \Sigma_{k=2}^{N} a_k z^k$. Prove that $\underline{\lim}\, R_S(N) \geq 1$ as $N \to \infty$. This implies that theorems about univalent polynomials can be used to prove theorems about S.

19. Let R_S be the radius of univalence of a polynomial $P(z)$. Prove that on the circle $|z| = R_S$, either there is one point ζ at which $P'(\zeta) = 0$, or there are two points ζ_1 and ζ_2 for which $P(\zeta_1) = P(\zeta_2)$.

20. Let

$$\phi(x, \theta) \equiv 1 + a_2 x \frac{\sin 2\theta}{\sin \theta} + a_3 x^2 \frac{\sin 3\theta}{\sin \theta} + \cdots + a_N x^{N-1} \frac{\sin N\theta}{\sin \theta}, \tag{38}$$

where x is a complex variable. Prove that for the polynomial $P_N(z)$ of Problem 18 the radius of univalence is the radius of the largest circle with center at the origin that contains no root in x of $\phi(x, \theta) = 0$ for all θ in $[0, \pi]$. Hint: In Problem 19 set $\zeta_k = e^{i\theta_k}$ $(k = 1, 2)$, $x = Re^{i(\theta_1+\theta_2)/2}$, and $\theta = (\theta_1 - \theta_2)/2$. The function $\phi(x, \theta)$ is called the *Dieudonné polynomial* for $P_N(z)$, see Dieudonné 1931 [1-B-242, p. 310]. For some applications of this polynomial see Brannan 1967 [2-B-171] and 1970 [2-B-174], and Rahman and Szynal 1978 [3-B-736]. For an interesting extension see Suffridge 1972 [2-B-1406].

21. Let $P_n(z) = z + \Sigma_{k=2}^{n} a_k z^k$ be a polynomial that is univalent in E. Prove that $|a_n| \leq 1/n$. For bounds on the other coefficients see D. Horowitz 1977 [3-B-367].

22. Suppose that in Problem 21 we have $a_n = 1/n$. Prove that all the zeros of $P'(z)$ lie on $|z| = 1$. A polynomial with this property is called a *C-polynomial*. For some facts about C-polynomials and recent results see Saff and Sheil-Small 1974 [3-B-780].

23. Prove that if $C(z) = 1 + \Sigma_1^{n-1} c_k z^k$ is a C-polynomial and $c_{n-1} = 1$, then $c_k = \bar{c}_{n-k-1}$ for $k = 0, 1, 2, \ldots, n-1$. Hint: Consider the roots of the related polynomial $C^*(z) \equiv z^{n-1}\overline{C(1/\bar{z})} = 1 + \Sigma_1^{n-1} \bar{c}_{n-1-k} z^k$.

24. Suppose that $P(z)$ in Problem 21 is in S and $a_n = 1/n$. Prove that $(k+1)a_k = (n-k)\bar{a}_{n-k}$ for $k = 1, 2, \ldots, n-1$. Brannan 1967 [2-B-171].

★ **25.** Let $\mathscr{F}$ be the family of functions $f(z) = z + a_2z^2 + \cdots$ that are regular and assume no value more than twice in E (the equation $f(z) = w_0$ has at most two roots in E for each w_0). Prove that the functional $J[f] = |a_3|$ is not a bounded functional in $\mathscr{F}$. Goodman 1948 [1-B-428].

CHAPTER 5

SOME THEOREMS ON POWER SERIES

1. COEFFICIENTS FOR THE AREA THEOREM

In proving that $|a_2| \leq 2$ for the class S we saw that the transformation

$$(1) \qquad \phi(\zeta) = \frac{1}{[f(1/\zeta^2)]^{1/2}} = \zeta + c_1 \frac{1}{\zeta} + c_3 \frac{1}{\zeta^3} + \cdots$$

was very useful when combined with the area theorem. Thus it is natural to ask if the other coefficients on the right side of (1) also give useful results. Further (as in the proof of Theorem 4.3), one may wish to consider $[f(z^k)]^{1/k}$ for some positive integer $k \neq 2$. Certainly such considerations yield new inequalities but because of the complexity of the coefficients c_n as functions of $a_2, a_3, \ldots$ the method is not as fruitful as one might hope.

In this chapter we interrupt our study of univalent functions to develop some formulas for the coefficients of power series. The impatient reader may omit this chapter and proceed to Chapter 6, returing to this chapter only when he or she feels the need.

Our first result gives the coefficients in (1) when the exponents 2 and $1/2$ are replaced by k and $1/k$, respectively.

Theorem 1. PWO. Let $k \neq 0$ be an integer and suppose that

$$f(z) = z + \sum_{n=2}^{\infty} a_n z^n \tag{2}$$

has no zeros[1] in E, except the one at the origin. Set

$$\phi(\zeta) \equiv \frac{1}{[f(1/\zeta^k)]^{1/k}} = \zeta + \sum_{n=1}^{\infty} \frac{c_{nk-1}}{\zeta^{nk-1}}. \tag{3}$$

Set

$$\gamma(k, m) \equiv (k+1)(2k+1)(3k+1) \cdots (mk+1) = \prod_{j=1}^{m} (jk+1),$$

with the usual convention that $\gamma(k, 0) = 1$. Then

$$c_{nk-1} = \sum_{m=1}^{n} \frac{(-1)^m \gamma(k, m-1)}{k^m} \left(\sum_{\sigma_m(n)} \frac{a_2^{r_1} a_3^{r_2} \cdots a_{n+1}^{r_n}}{r_1! r_2! \cdots r_n!} \right), \tag{4}$$

where the second sum is taken over all n-tuples in $\sigma_m(n)$, and $\sigma_m(n)$ is the collection of all n-tuples $(r_1, r_2, \ldots, r_n)$ of nonnegative integers for which

$$r_1 + 2r_2 + 3r_3 + \cdots + nr_n = n \tag{5}$$

and

$$r_1 + r_2 + r_3 + \cdots + r_n = m. \tag{6}$$

[1] In these theorems univalence has nothing to do with the relations among the coefficients. The hypothesis of no zeros merely guarantees the domain of convergence. Henceforth we will omit such conditions. We trust that the reader can decide the domain of convergence of the series we obtain.

This theorem is merely a reformulation of the multinomial theorem applied to an infinite series. It is stated in a form that is convenient for the area theorem. A proof that is independent of the multinomial theorem can be found in Goodman 1972 [2-B-481]. For a different proof, see Todorov 1979 [3-B-927].

The first six cases of this theorem ($n = 2, 3, \ldots, 7$) give the formulas

$$\text{(7)}\quad c_{k-1} = -\frac{1}{k}a_2,$$

$$\text{(8)}\quad c_{2k-1} = -\frac{1}{k}a_3 + \frac{k+1}{2k^2}a_2^2,$$

$$\text{(9)}\quad c_{3k-1} = -\frac{1}{k}a_4 + \frac{k+1}{k^2}a_2a_3 - \frac{(2k+1)(k+1)}{6k^3}a_2^3,$$

$$\text{(10)}\quad c_{4k-1} = -\frac{1}{k}a_5 + \frac{k+1}{k^2}a_2a_4 + \frac{k+1}{2k^2}a_3^2 - \frac{(2k+1)(k+1)}{2k^3}a_2^2a_3 + \frac{(3k+1)(2k+1)(k+1)}{24k^4}a_2^4,$$

$$\text{(11)}\quad c_{5k-1} = -\frac{1}{k}a_6 + \frac{k+1}{k^2}a_2a_5 + \frac{k+1}{k^2}a_3a_4 - \frac{\gamma(k,2)}{2k^3}(a_2^2a_4 + a_2a_3^2) + \frac{\gamma(k,3)}{6k^4}a_2^3a_3 - \frac{\gamma(k,4)}{120k^5}a_2^5,$$

$$\text{(12)}\quad c_{6k-1} = -\frac{1}{k}a_7 + \frac{k+1}{2k^2}(2a_2a_6 + 2a_3a_5 + a_4^2) - \frac{\gamma(k,2)}{6k^3}(3a_2^2a_5 + 6a_2a_3a_4 + a_3^3) + \frac{\gamma(k,3)}{12k^4}(2a_2^3a_4 + 3a_2^2a_3^2) - \frac{\gamma(k,4)}{24k^5}a_2^4a_3 + \frac{\gamma(k,5)}{720k^6}a_2^6.$$

We can express the coefficients c_{nk-1} as determinants. Indeed, if we differentiate (3) we obtain

$$\text{(13)}\quad \frac{-1}{k}\,\frac{f'(1/\zeta^k)}{[f(1/\zeta^k)]^{1+1/k}}\,\frac{(-k)}{\zeta^{k+1}} = 1 + \sum_{n=1}^{\infty}\frac{(1-nk)c_{nk-1}}{\zeta^{nk}}.$$

In the left side we use the identities,

$$\text{(14)}\quad \frac{1}{\zeta[f(1/\zeta^k)]^{1/k}} = 1 + \sum_{n=1}^{\infty}\frac{c_{nk-1}}{(\zeta^k)^n}$$

and

$$\frac{1}{\zeta^k} f'\left(\frac{1}{\zeta^k}\right) = \frac{1}{\zeta^k} + \sum_{n=1}^{\infty} \frac{na_n}{(\zeta^k)^n}. \tag{15}$$

We replace $1/\zeta^k$ by z and make the usual agreement that $a_1 = 1$ and $c_{-1} = 1$. Then (13) gives

$$\left(\sum_{n=1}^{\infty} na_n z^n\right)\left(\sum_{n=0}^{\infty} c_{nk-1} z^n\right) = \left(\sum_{n=1}^{\infty} a_n z^n\right)\left(\sum_{n=0}^{\infty} (1 - nk)c_{nk-1} z^n\right). \tag{16}$$

With $n + 1$ fixed, we equate the coefficients of z^{n+1} on both sides of (16). This yields

$$\sum_{j=0}^{n} (jk + n - j)a_{n+1-j}c_{jk-1} = 0, \qquad n = 1, 2, 3, \ldots. \tag{17}$$

We adjoin to this system of equations the trival equation $c_{-1} = 1$. Then the first $n + 1$ equations are linear in the variables $c_{-1}, c_{k-1}, c_{2k-1}, \ldots, c_{nk-1}$. By Cramer's rule we find that

$$c_{nk-1} = \frac{(-1)^n}{n!k^n} \begin{vmatrix} a_2 & k & 0 & \cdots & 0 \\ 2a_3 & (k+1)a_2 & 2k & \cdots & 0 \\ 3a_4 & (k+2)a_3 & (2k+1)a_2 & \cdots & 0 \\ \cdot & \cdot & \cdot & \cdots & \cdot \\ \cdot & \cdot & \cdot & \cdots & \cdot \\ \cdot & \cdot & \cdot & \cdots & (n-1)k \\ na_{n+1} & (k+n-1)a_n & (2k+n-2)a_{n-1} & \cdots & [(n-1)k+1]a_2 \end{vmatrix}. \tag{18}$$

Thus we have proved[2]

Theorem 2. Jacobsthal [1]. If $k \neq 0$ is a positive integer and c_{nk-1} is defined by equation (3), then c_{nk-1} is given by the determinant (18). In (18) A_{ij} (the term in the ith row and jth column) is given by

$$A_{ij} = \begin{cases} [i + (j-1)(k-1)]a_{2+i-j}, & \text{if } i + 1 \geq j, \\ 0, & \text{if } i + 1 < j, \end{cases} \tag{19}$$

and $a_1 = c_{-1} = 1$.

[2] I am indebted to Prof. H. Waadeland for calling my attention to this reference.

Since both (4) and (18) give the same quantities, it is clear that (4) represents the (unique) evaluation of the determinant (18).

If we replace k by $-K$, we obtain a formula for the coefficients in

(20) $$g(z) \equiv [f(z^K)]^{1/K} = z + \sum_{n=1}^{\infty} b_{nK+1} z^{nK+1}.$$

Thus b_{nK+1} is given by equation (4) or (18) when k is replaced by $-K$, and c_{nk-1} is replaced by b_{nK+1}.

One may ask if the formula given in Theorem 1 has meaning if k is not an integer. As it stands, it has no meaning, but with a little modification it is meaningful and turns out to be useful in connection with an inequality due to Prawitz that will be developed in Chapter 16.3.

We observe that if $f(z)$ is in S, then $f(z)/z$ has no zeros in E. Hence for any real α we can consider the function

(21) $$h(z) \equiv \left[\frac{f(z)}{z}\right]^{-\alpha/2} = \left[\frac{z}{f(z)}\right]^{\alpha/2} \equiv 1 + \sum_{n=1}^{\infty} b_n z^n,$$

where the expansion on the right side converges in E. As a purely formal operation, we divide both sides by $z^{\alpha/2}$ (a meaningless quantity) and compare the result

(22) $$\frac{1}{[f(z)]^{\alpha/2}} = \frac{1}{z^{\alpha/2}} + \sum_{n=1}^{\infty} b_n z^{n-\alpha/2}$$

with that obtained from (3) when $1/\zeta^k$ is replaced by z:

(23) $$\frac{1}{[f(z)]^{1/k}} = \frac{1}{z^{1/k}} + \sum_{n=1}^{\infty} c_{nk-1} z^{n-1/k}.$$

It is clear that if we set $1/k = \alpha/2$, the series (22) and (23) are identical. With a little effort those steps can be justified to give

Theorem 3. PWO. If $f(z)$ is in S and $\alpha \neq 0$ is any real number, then the coefficients b_n in the series

(21) $$\left[\frac{z}{f(z)}\right]^{\alpha/2} = 1 + \sum_{n=1}^{\infty} b_n z^n$$

are given by

(24) $$b_n = c_{nk-1}, \qquad k = 2/\alpha,$$

where c_{nk-1} is given by either of the formulas in Theorems 1 and 2.

2. COEFFICIENTS FOR THE INVERSE OF A FUNCTION

It is possible to give a formula for the coefficients in the Maclaurin series for the inverse of a function. As usual we suppose that

$$w = f(z) = z + \sum_{n=2}^{\infty} a_n z^n \tag{2}$$

and we let

$$z = g(w) = w + \sum_{n=2}^{\infty} b_n w^n \tag{25}$$

be the inverse function, which will always exist in some small disk with center at $w = 0$. Our object is to find b_n as a function of $a_2, a_3, \ldots, a_n$. This problem is partially solved as a special case of a more general theorem, in which an arbitrary function $F(z)$ is expressed as a power series in w.

Theorem 4. Let $F(z)$ be regular in a disk that contains $z = 0$, and let $f(z)$ and $g(w)$, given by equations (2) and (25), be inverse functions in the same disk. Let

$$h(z) = \frac{f(z)}{z} = 1 + a_2 z + a_3 z^2 + \cdots. \tag{26}$$

Then in some neighborhood of $z = 0$

$$F(z) = F(0) + \sum_{n=1}^{\infty} \frac{w^n}{n!} \left[\frac{d^{n-1}}{dz^{n-1}} \left(\frac{F'(z)}{[h(z)]^n} \right) \right] \Bigg|_{z=0}. \tag{27}$$

The series in (27) is called the Bürmann-Lagrange series (see Pólya-Szegö [1-B-1064, vol. I, p. 125]).

Proof. Let C be a suitably small circle about the origin in the z-plane and let Γ be its image under $f(z)$ in the w-plane. Here we select C so small that Γ is a simple closed curve that encloses the point $w = 0$. If

$$F(z) \equiv F(g(w)) = \sum_{n=0}^{\infty} c_n w^n, \tag{28}$$

then by the Cauchy formula

(29) $$c_n = \frac{1}{2\pi i}\int_\Gamma \frac{F(z)}{w^{n+1}}\,dw.$$

We transfer the path of integration from Γ to C by changing the variable of integration from w to z. Then (29) becomes

(30) $$c_n = \frac{1}{2\pi i}\int_C \frac{F(z)f'(z)}{[f(z)]^{n+1}}\,dz.$$

An integration by parts with $U = F(z)$ and $dV = [f(z)]^{-(n+1)}f'(z)\,dz$ $(n \neq 0)$ gives

$$c_n = \frac{1}{2\pi i}\Delta_C \frac{F(z)}{-n[f(z)]^n} + \frac{1}{2\pi i}\int_C \frac{F'(z)}{n[f(z)]^n}\,dz,$$

and since the first term is 0,

(31) $$c_n = \frac{1}{2\pi i}\int_C \frac{F'(z)}{n[f(z)]^n}\,dz = \frac{1}{2\pi i}\int_C \frac{F'(z)}{nz^n[h(z)]^n}\,dz.$$

On the other hand the right side in (31) is the coefficient of z^{n-1} in the Maclaurin series for $F'(z)/n[h(z)]^n$. Consequently

$$c_n = \frac{1}{(n-1)!}\,\frac{d^{n-1}}{dz^{n-1}}\left[\frac{F'(z)}{n[h(z)]^n}\right]\Bigg|_{z=0}$$

and this gives (27). ■

If we set $F(z) \equiv z$, then we have as a corollary

Theorem 5. If $f(z)$ and $g(w)$ are inverse functions, where

(2) $$w = f(z) = z + \sum_{n=2}^{\infty} a_n z^n$$

and

(25) $$z = g(w) = w + \sum_{n=2}^{\infty} b_n w^n,$$

then

(32) $$b_n = \frac{1}{n!}\,\frac{d^{n-1}}{dz^{n-1}}[h(z)]^{-n}\Bigg|_{z=0}$$

where $h(z) = 1 + \Sigma_2^\infty a_n z^{n-1}$.

The formula in Theorem 5 is attractive looking, but unless $h(z)$ is carefully selected, the computation of b_n can be quite laborious. We illustrate the term "carefully selected" in

Example 1. Find a Maclaurin series for the inverse function if $w = ze^{-Az}$. Find the radius of convergence of the series.

Solution. Here $h(z) = e^{-Az}$. Hence

$$b_n = \frac{1}{n!} \frac{d^{n-1}}{dz^{n-1}}(e^{Anz})\bigg|_{z=0} = \frac{A^{n-1}n^{n-1}}{n!},$$

and

$$z = \sum_{n=1}^{\infty} \frac{A^{n-1}n^{n-1}}{n!} w^n. \tag{33}$$

The ratio test shows that the series (33) converges for $|w| < 1/|A|e$.

The Bürmann-Lagrange series has a variety of useful applications. The interested reader should consult Pólya-Szegö [1-B-1064, vol. 1, p. 125], and Frame 1957 [1-B-323].

Theorem 6. PWO. If $f(z)$ and $g(w)$, given by equations (2) and (25), are inverse functions, then for $n \geq 2$

$$b_n = \sum_{m=1}^{n-1} (-1)^m \frac{(n+m-1)!}{n!} \left(\sum_{\sigma_m(n-1)} \frac{a_2^{r_1} a_3^{r_2} \cdots a_n^{r_{n-1}}}{r_1! r_2! \cdots r_{n-1}!} \right), \tag{34}$$

where the second sum is over all $(n-1)$-tuples $(r_1, r_2, \ldots, r_{n-1})$ of nonnegative integers in $\sigma_m(n-1)$, and $\sigma_m(n-1)$ is the set of those $(n-1)$-tuples for which

$$r_1 + 2r_2 + \cdots + (n-1)r_{n-1} = n - 1 \tag{35}$$

and

$$r_1 + r_2 + \cdots + r_{n-1} = m. \tag{36}$$

Theorem 7. PWO. Under the conditions of Theorem 6

$$
\tag{37}
b_n = \frac{(-1)^{n+1}}{n!}
\begin{vmatrix}
na_2 & 1 & 0 & \cdots & 0 \\
2na_3 & (n+1)a_2 & 2 & \cdots & 0 \\
3na_4 & (2n+1)a_3 & (n+2)a_2 & \cdots & 0 \\
\cdot & \cdot & \cdot & \cdot & \cdot \\
\cdot & \cdot & \cdot & \cdot & \cdot \\
\cdot & \cdot & \cdot & \cdot & \cdot \\
\cdot & \cdot & \cdot & \cdots & n-2 \\
(n-1)na_n & [(n-2)n+1]a_{n-1} & [(n-3)n+2]a_{n-2} & \cdots & (2n-2)a_2
\end{vmatrix}
$$

The elements in the determinants $|A_{ij}|$ in (37) are given by

$$
\tag{38}
A_{ij} = \begin{cases} [(i-j+1)n+j-1]a_{i-j+2}, & \text{if } i+1 \geq j, \\ 0, & \text{if } i+1 < j. \end{cases}
$$

Some particular examples of (37) are: $b_2 = -a_2$,

$$
\tag{39}
b_3 = \frac{(-1)^4}{3!} \begin{vmatrix} 3a_2 & 1 \\ 6a_3 & 4a_2 \end{vmatrix} = 2a_2^2 - a_3,
$$

$$
\tag{40}
b_4 = \frac{(-1)^5}{4!} \begin{vmatrix} 4a_2 & 1 & 0 \\ 8a_3 & 5a_2 & 2 \\ 12a_4 & 9a_3 & 6a_2 \end{vmatrix} = 5a_2a_3 - 5a_2^3 - a_4,
$$

$$
\tag{41}
b_5 = \frac{(-1)^6}{5!} \begin{vmatrix} 5a_2 & 1 & 0 & 0 \\ 10a_3 & 6a_2 & 2 & 0 \\ 15a_4 & 11a_3 & 7a_2 & 3 \\ 20a_5 & 16a_4 & 12a_3 & 8a_2 \end{vmatrix}
$$

$$
= 6a_2a_4 - 21a_2^2a_3 + 14a_2^4 + 3a_3^2 - a_5.
$$

Theorems 6 and 7 are simplified versions of two formulas proved by Kamber [1] who considered a much more complicated relation between $f(z)$ and $g(w)$.

Todorov 1981 [3-B-931] has developed explicit formulas for $f^{(n)}(z)$ when $f(z) = g(h(z))$ is a composite function. Theorems 6 and 7 are special cases of his formulas. For further work on this type problem, see the references cited by Todorov.

3. TRANSFORMATION OF THE RANGE FROM THE UNIT DISK TO THE RIGHT HALF-PLANE

There are two more series transformations that are occasionally useful and are much simpler than the ones treated so far.

Definiton 1. Let B_0 denote the set of all functions of the form

$$f(z) = \sum_{n=1}^{\infty} b_n z^n = b_1 z + \cdots \tag{42}$$

that are regular in E and for which $|f(z)| < 1$ in E.

The set B_0 is often referred to briefly as the set of bounded functions. These are the functions that satisfy the conditions of Schwarz's Lemma (see Theorem 7.8). The normalization $b_1 = 1$ is not available, because if $f(z)$ is in B_0, then by Schwarz's Lemma either $|b_1| < 1$ or $f(z) \equiv e^{i\alpha} z$.

Definition 2. Let P denote the set of all functions of the form

$$g(z) = 1 + \sum_{n=1}^{\infty} p_n z^n \tag{43}$$

that are regular in E and for which $\operatorname{Re} g(z) > 0$ in E.

Since the Möbius transformations

$$w = \frac{1+z}{1-z} \quad \text{and} \quad z = \frac{w-1}{w+1} \tag{44}$$

carry $|z| < 1$ onto $\operatorname{Re} w > 0$ and back, the classes B_0 and P are related by the equations

$$g(z) = \frac{1+f(z)}{1-f(z)} \quad \text{and} \quad f(z) = \frac{g(z)-1}{g(z)+1}. \tag{45}$$

Indeed, for either of these equations, $f(z)$ is in B_0 if and only if $g(z)$ is in P. The relation among the coefficients is given in

Theorem 8. PLE. Let $f(z)$ and $g(z)$ have the Maclaurin series given in equations (42) and (43), respectively. If $f(z)$ and $g(z)$ are related by equation (45), then for $n \geq 1$

(46) $$p_n = 2\begin{vmatrix} b_1 & -1 & 0 & 0 & \cdots & 0 \\ b_2 & b_1 & -1 & 0 & \cdots & 0 \\ b_3 & b_2 & b_1 & -1 & \cdots & 0 \\ \cdot & \cdot & \cdot & \cdot & \cdot & \cdot \\ \cdot & \cdot & \cdot & \cdot & \cdot & \cdot \\ \cdot & \cdot & \cdot & \cdot & \cdot & \cdot \\ b_{n-1} & b_{n-2} & b_{n-3} & b_{n-4} & \cdots & -1 \\ b_n & b_{n-1} & b_{n-2} & b_{n-3} & \cdots & b_1 \end{vmatrix},$$

and

(47) $$b_n = \frac{(-1)^{n+1}}{2^n}\begin{vmatrix} p_1 & 2 & 0 & \cdots & 0 & 0 \\ p_2 & p_1 & 2 & \cdots & 0 & 0 \\ p_3 & p_2 & p_1 & \cdots & 0 & 0 \\ \cdot & \cdot & \cdot & \cdot & \cdot & \cdot \\ \cdot & \cdot & \cdot & \cdot & \cdot & \cdot \\ \cdot & \cdot & \cdot & \cdot & \cdot & \cdot \\ p_{n-1} & p_{n-2} & p_{n-3} & \cdots & p_1 & 2 \\ p_n & p_{n-1} & p_{n-2} & \cdots & p_2 & p_1 \end{vmatrix},$$

Some particular examples of (46) and (47) are:

(48) $$\begin{cases} p_1 = 2b_1, \qquad p_2 = 2b_2 + 2b_1^2 \\ p_3 = 2b_3 + 4b_1b_2 + 2b_1^3, \\ p_4 = 2b_4 + 4b_1b_3 + 2b_2^2 + 6b_1^2b_2 + 2b_1^4, \end{cases}$$

and

(49) $$\begin{cases} b_1 = \dfrac{1}{2}p_1, \qquad b_2 = \dfrac{1}{4}(2p_2 - p_1^2), \\ b_3 = \dfrac{1}{8}(4p_3 - 4p_1p_2 + p_1^3), \\ b_4 = \dfrac{1}{16}(8p_4 - 8p_1p_3 - 4p_2^2 + 6p_1^2p_2 - p_1^4). \end{cases}$$

PROBLEMS

1. Show that Theorem 1 gives the formulas for c_{nk-1} as stated in equations (7), (8), . . . , (12).

★ 2. Without using Theorem 1, derive equations (7), (8), and (9) directly from equation (3).

3. Use the determinant, equation (18), with $n = 1$, 2, and 3 to derive formulas (7), (8), and (9).

4. Let b_k be defined by the expansion in equation (21). Use the process described in Theorem 3 to obtain the following formulas for b_1, b_2, b_3, and b_4:

$$
\begin{aligned}
b_1 &= -\alpha \frac{a_2}{2}, \\
b_2 &= -\alpha \frac{a_3}{2} + \alpha(\alpha+2)\frac{a_2^2}{8}, \\
b_3 &= -\alpha \frac{a_4}{2} + \alpha(\alpha+2)\frac{a_2 a_3}{4} - \alpha(\alpha+2)(\alpha+4)\frac{a_2^3}{48}, \\
b_4 &= -\alpha \frac{a_5}{2} + \alpha(\alpha+2)\frac{a_2 a_4}{4} + \alpha(\alpha+2)\frac{a_3^2}{8} \\
&\quad - \alpha(\alpha+2)(\alpha+4)\frac{a_2^2 a_3}{16} + \alpha(\alpha+2)(\alpha+4)(\alpha+6)\frac{a_2^4}{384}.
\end{aligned} \tag{50}
$$

Bernardi 1952 [1-B-103] used these formulas to prove that if $f(z)$ is in S, then $|a_4| < 4.0891$.

5. It is easy to see that the inverse of $w = z/(1-z)$ is $z = w/(1+w)$. Use the formula of Theorem 5 to obtain the coefficients of the Maclaurin series for $w/(1+w)$.

6. Use Theorem 5 to find the Maclaurin series for the inverse of the Koebe function.

7. We can use Theorem 5 to solve equations as indicated by the following example. To solve the fifth degree equation $z^5 - 3z + 1 = 0$ we put this equation in the form

$$\frac{1}{3} = \frac{z}{1+z^5}.$$

We replace $1/3$ by w and express z as a power series in w. Then we obtain a root of $z^5 - 3z + 1 = 0$ by setting $w = 1/3$ in the series. Find the infinite series for this problem. Hint: $[h(z)]^{-n} = (1+z^5)^n$.

8. If $a_0 \neq 0$ set

$$\sum_{n=0}^{\infty} b_n z^n = \frac{1}{\sum_{n=0}^{\infty} a_n z^n}.$$

Prove that $b_0 = 1/a_0$, $b_1 = -a_1/a_0^2$, and for $n > 1$,

$$\text{(51)} \qquad b_n = \frac{(-1)^n}{a_0^{n+1}} \begin{vmatrix} a_1 & a_0 & 0 & \cdots & 0 \\ a_2 & a_1 & a_0 & \cdots & 0 \\ \cdot & \cdot & \cdot & \cdot & \cdot \\ \cdot & \cdot & \cdot & \cdot & \cdot \\ \cdot & \cdot & \cdot & \cdot & \cdot \\ a_{n-1} & a_{n-2} & a_{n-3} & \cdots & a_0 \\ a_n & a_{n-1} & a_{n-2} & \cdots & a_1 \end{vmatrix}$$

9. Prove Theorem 8.

10. Check the equations in the set (48) by evaluating the determinant (46) for $n = 1, 2, 3$, and 4.

11. Check the equations in the set (48) by writing equation (45) in the form

$$g(z) = 1 + 2f(z) + 2f^2(z) + 2f^3(z) + \cdots$$

and performing the indicated operations on the right side.

12. Check the equations in the set (49) in two different ways.

CHAPTER 6

IMPLICATIONS OF THE BOUND ON THE SECOND COEFFICIENT

1. DISTORTION THEOREMS

The fact that $|a_2| \leq 2$ for any $f(z)$ in S can be used to prove other theorems about the class. Indeed, any transformation that carries one function of S into another function in S will give some expression for the second coefficient to which we can apply the bound just derived. If the transformation is selected properly, we obtain new results that are both interesting and sharp.

Theorem 1. If $f(z)$ is in S and $f(z)$ omits γ, then $|\gamma| \geq 1/4$. The inequality is sharp[1] and the equality sign occurs if and only if $f(z)$ is some rotation of $k(z) = z/(1-z)^2$.

[1] The Theory of Univalent Functions was initiated by P. Koebe 1907 [1-B-688], who proved the existence of a positive constant b such that $|\gamma| \geq b$. The exact value $b = 1/4$ also follows from Theorem 8, so credit for this result belongs to Plemelj and Pick. Both Theorem 1 and Theorem 8 depend heavily on the inequality $|a_2| \leq 2$ for $f(z)$ in S.

This theorem states that if $f(z)$ is in S, then the image of E under $f(z)$ must cover an open disk with center at the origin and radius 1/4. We refer to this theorem as a *distortion theorem*, because it tells us that in the class S there is a limit to the distortion of the boundary: it cannot be "pushed toward the origin" beyond the limit described in the theorem.

Proof. The transformation $w = W/(1 - W/\gamma)$ is univalent in any domain in the W-plane and is regular if the domain does not contain γ. Thus, if $W = f(z)$ and $f(z)$ omits γ, then (replacing W by $f(z)$) we see that

$$\text{(1)} \qquad g(z) \equiv \frac{f(z)}{1 - \dfrac{f(z)}{\gamma}} = f(z)\left[1 + \frac{f(z)}{\gamma} + \frac{f^2(z)}{\gamma^2} + \cdots\right]$$

is regular and univalent in E. Further (see Theorem 2.5),

$$\text{(2)} \qquad g(z) = z + \left(a_2 + \frac{1}{\gamma}\right)z^2 + \cdots$$

and hence $g(z)$ is in S. Consequently $|a_2 + 1/\gamma| \leq 2$. The triangle inequality yields

$$\text{(3)} \qquad \left|\frac{1}{\gamma}\right| - |a_2| \leq \left|a_2 + \frac{1}{\gamma}\right| \leq 2,$$

or $|1/\gamma| \leq 2 + |a_2| \leq 4$. This gives $|\gamma| \geq 1/4$. If $|\gamma| = 1/4$, then $|a_2| = 2$, and hence $f(z)$ is some rotation of the Koebe function. ■

Theorem 2. PLE. If $f(z)$ is in S and $|a_2| \neq 2$, then $f(E)$ covers the open disk

$$\text{(4)} \qquad \left|w - \frac{\overline{a_2}}{4 - |a_2|^2}\right| < \frac{2}{4 - |a_2|^2}.$$

Our next objective is a bound for $|f'(z)|$.

Lemma 1. If $f(z)$ is in S, then for any ζ in E

$$\text{(5)} \qquad \frac{1}{2}\left|\frac{f''(\zeta)}{f'(\zeta)}(1 - |\zeta|^2) - 2\bar{\zeta}\right| \leq 2$$

Remark. We have replaced the usual z by $\zeta = \varrho e^{i\theta}$ because we anticipate integrating this inequality with ϱ as the variable, $0 \le \varrho \le r$.

Proof. For each fixed λ in E, the function

$$g(z) = \frac{f\left(\dfrac{z+\lambda}{1+\bar{\lambda}z}\right) - f(\lambda)}{f'(\lambda)(1-|\lambda|^2)} = z + b_2 z^2 + \cdots \tag{6}$$

is in S. By Theorems 2.4 and 4.1 we have

$$|b_2| = \left| \frac{f''(\lambda)(1-|\lambda|^2)}{2f'(\lambda)} - \bar{\lambda} \right| \le 2. \tag{7}$$

Since λ is an arbitrary point in E, we may replace λ by ζ in (7). Then (7) gives the inequality (5). ■

Lemma 2. If $f(z)$ is regular at $\zeta = \varrho e^{i\theta}$ and $f'(\zeta) \neq 0$, then

$$\varrho \frac{\partial}{\partial \varrho} \ln|f'(\zeta)| = \operatorname{Re} \frac{\zeta f''(\zeta)}{f'(\zeta)}, \tag{8}$$

and[1]

$$\varrho \frac{\partial}{\partial \varrho} \arg f'(\zeta) = \operatorname{Im} \frac{\zeta f''(\zeta)}{f'(\zeta)}. \tag{9}$$

Proof. Consider

$$\varrho \frac{\partial}{\partial \varrho} \ln f'(\zeta) = \varrho \frac{\partial}{\partial \varrho} (\ln|f'(\zeta)| + i \arg f'(\zeta)). \tag{10}$$

For the left side of (10) we have

$$\begin{aligned} \varrho \frac{\partial}{\partial \varrho} \ln f'(\zeta) &= \varrho \frac{d}{d\zeta} (\ln f'(\zeta)) \frac{\partial \zeta}{\partial \varrho} = \varrho \frac{f''(\zeta)}{f'(\zeta)} e^{i\theta} \\ &= \zeta \frac{f''(\zeta)}{f'(\zeta)}. \end{aligned} \tag{11}$$

[1] Whenever we encounter the multiple-valued function $\arg f'(z)$, it is understood that some branch has been chosen that is continuous in the domain under consideration. If $f(z)$ is in S, we determine $\arg f'(z)$ by requiring that $\arg f'(0) = \arg 1 = 0$.

On the right sides of (10) and (11) we equate the real parts to obtain (8) and the imaginary parts to obtain (9). ∎

Theorem 3. The Distortion Theorem. If $f(z)$ is in S, then for each $z = re^{i\theta}$ in E

$$\frac{1-r}{(1+r)^3} \leq |f'(z)| \leq \frac{1+r}{(1-r)^3}, \tag{12}$$

and both sides of the inequality (12) are sharp. Equality occurs for the Koebe function $f(z) = z/(1-z)^2$.

Proof. We multiply (5) by $2|\zeta/(1-|\zeta|^2)|$ and obtain

$$\left| \zeta \frac{f''(\zeta)}{f'(\zeta)} - \frac{2|\zeta|^2}{1-|\zeta|^2} \right| \leq \frac{4|\zeta|}{1-|\zeta|^2}. \tag{13}$$

We replace $|\zeta|$ by ϱ. Then (13) yields the two inequalities

$$-\frac{4\varrho}{1-\varrho^2} \leq \operatorname{Re}\left(\frac{\zeta f''(\zeta)}{f'(\zeta)} - \frac{2\varrho^2}{1-\varrho^2}\right) \leq \frac{4\varrho}{1-\varrho^2} \tag{14}$$

and

$$-\frac{4\varrho}{1-\varrho^2} \leq \operatorname{Im}\left(\frac{\zeta f''(\zeta)}{f'(\zeta)} - \frac{2\varrho^2}{1-\varrho^2}\right) \leq \frac{4\varrho}{1-\varrho^2}. \tag{15}$$

We transpose the real term $2\varrho^2/(1-\varrho^2)$ and use Lemma 2. Then (14) and (15) yield

$$\frac{2\varrho^2 - 4\varrho}{1-\varrho^2} \leq \varrho \frac{\partial}{\partial \varrho} \ln |f'(\zeta)| \leq \frac{2\varrho^2 + 4\varrho}{1-\varrho^2} \tag{16}$$

and

$$-\frac{4\varrho}{1-\varrho^2} \leq \varrho \frac{\partial}{\partial \varrho} \arg f'(\zeta) \leq \frac{4\varrho}{1-\varrho^2}. \tag{17}$$

In (16) we divide by ϱ and then integrate along the straight line path from $\zeta = 0$ to $\zeta = re^{i\theta}$. Thus ϱ runs from 0 to r. Since $f'(0) = 1$, we obtain

$$\ln \frac{1-r}{(1+r)^3} \leq \ln |f'(z)| \leq \ln \frac{1+r}{(1-r)^3}. \tag{18}$$

But this inequality is equivalent to (12). To see that the inequality is sharp we observe that for the Koebe function

$$k'(z) = \frac{d}{dz}\frac{z}{(1-z)^2} = \frac{1+z}{(1-z)^3}. \tag{19}$$

We set $z = r$ for the right side of (12) and $z = -r$ for the left side of (12). ■

2. THE ARGUMENT OF THE DERIVATIVE OF A UNIVALENT FUNCTION

If we integrate the inequality (17), we obtain

Theorem 4. PLE. Bieberbach 1919 [1-B-118]. If $f(z)$ is in S, then for each z in E,

$$|\arg f'(z)| \le 2 \ln \frac{1+r}{1-r}. \tag{20}$$

The inequality (20) is not sharp. Using more complicated and deeper methods due to Loewner 1923 [1-B-882], the Russian mathematician G. M. Goluzin 1936 [1-B-385 and 386] obtained the sharp version:

Theorem 5. PWO. The Rotation Theorem. If $f(z)$ is in S, then

$$|\arg f'(z)| \le \begin{cases} 4 \arcsin r, & \text{if } r \le \sqrt{2}/2, \\ \pi + \ln \dfrac{r^2}{1-r^2}, & \text{if } \sqrt{2}/2 < r < 1. \end{cases} \tag{21}$$

Gutljanskii 1971 [2-B-531] found the sharp bound for $|\arg f'(z)|$ in the set $S^{(k)}$. In an appendix to the book by Schaeffer and Spencer, 1950 [1-B-1240], Grad proved a far-reaching generalization of Theorem 5. If z_0 is a fixed point in E, he found the set of all possible values of $\ln f'(z_0) = \ln |f'(z_0)| + i \arg f'(z_0)$ as $f(z)$ ranges over all functions in S.

We observe that the inequalities (13) and (14) will be useful later. Hence we state the results (after transposing the term $2\varrho/(1-\varrho^2)$ and changing ϱ to r) as

Theorem 6. If $f(z)$ is in S, then for any $z = re^{i\theta}$ in E

(22)
$$\left| z \frac{f''(z)}{f'(z)} \right| \leq \frac{2r^2 + 4r}{1 - r^2}$$

and

(23)
$$\frac{2r^2 - 4r}{1 - r^2} \leq \operatorname{Re} \frac{zf''(z)}{f'(z)} \leq \frac{2r^2 + 4r}{1 - r^2}.$$

In Chapter 8 we will see that on the circle $z = re^{i\theta}$

$$\operatorname{Re} z \frac{f''(z)}{f'(z)} = \frac{d}{d\theta} (\arg f'(z)).$$

Consequently the double inequality (23) gives upper and lower bounds for the rate of change of $\arg f'(z)$ on that circle.

3. THE MODULUS OF A UNIVALENT FUNCTION

We may expect that if we integrate the inequality (12), we will obtain bounds for $|f(z)|$. Privalov 1924 [1-B-1083, p. 357] generalized this idea to obtain.

Theorem 7. Suppose that $f(z)$ is in S and that for $0 \leq r < 1$. we have

(24)
$$m'(r) \leq |f'(z)| \leq M'(r)$$

where $m'(r)$ and $M'(r)$ are real-valued functions of r in $[0, 1)$. Then

(25)
$$\int_0^r m'(t)\, dt \leq |f(z)| \leq \int_0^r M'(t)\, dt.$$

Proof. The right side of (25) is trivial. Indeed, integrating along a radial line $\zeta = te^{i\alpha}$

(26)
$$|f(re^{i\theta})| = \left| \int_0^z f'(\zeta)\, d\zeta \right| = \left| \int_0^r f'(te^{i\alpha}) e^{i\alpha}\, dt \right|$$
$$\leq \int_0^r |f'(te^{i\alpha})|\, dt \leq \int_0^r M'(t)\, dt.$$

The left side of (25) requires more care. Let $\Gamma = f(C_r)$ and let w_1 be some point of Γ that is nearest to the origin. By a rotation we may assume that $w_1 > 0$ (see Figure 1 next page).

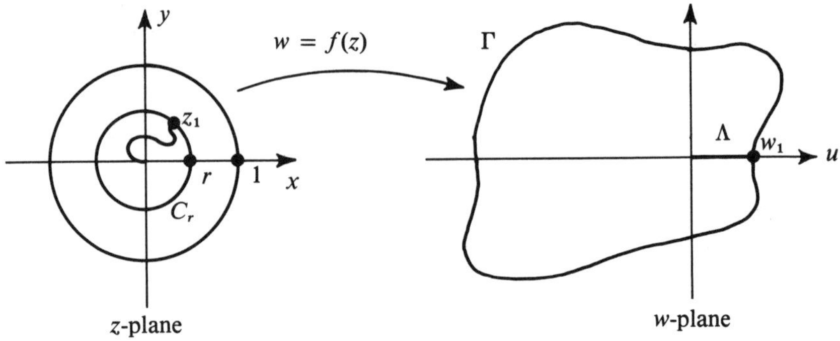

FIGURE 1.

Let Λ be the line segment $0 \leq w \leq w_1$ and suppose that z_1 and the curve L are the preimages of w_1 and Λ, respectively, under $f(z)$. With ζ as the variable of integration on L we have on L that $f'(\zeta)\,d\zeta = dw > 0$. Hence

$$w_1 = \int_0^{w_1} dw = \int_0^{z_1} f'(\zeta)\,d\zeta = \int_0^{z_1} |f'(\zeta)|\,|d\zeta| \tag{27}$$

$$\geq \int_0^r |f'(te^{i\theta})|\,dt \geq \int_0^r m'(t)\,dt. \qquad \blacksquare$$

Theorem 8. If $f(z)$ is in S, then for any z in E

$$\frac{r}{(1+r)^2} \leq |f(z)| \leq \frac{r}{(1-r)^2}, \qquad r = |z|. \tag{28}$$

Proof. By Theorem 3 we may set $m'(r) = (1-r)/(1+r)^3$ and $M'(r) = (1+r)/(1-r)^3$ in Theorem 7. Integration will give (28). ■

A review of Sections 1, 2, and 3 shows that the proof of the central items (Theorems 3 and 8) depended heavily on the fact that $|a_2| \leq 2$. Suppose that we know only that $|a_2| \leq K$, and this was the situation before 1916. In 1913 Plemelj [1-B-1057] announced that he had obtained upper and lower bounds for both $|f'(z)|$ and $|f(z)|$ and his bounds were expressed as functions of K. When we set $K = 2$ in Plemelj's formulas, we obtain the sharp bounds given in Theorems 3 and 8. Unfortunately, Plemelj in his paper gives no indication of his method of proof. This work was followed by Pick 1916 [1-B-1048] who obtained all of Plemelj's formulas with the same unknown parameter $K = \max |a_2|$, but Pick's proofs were rather complicated. The beautiful organization of this material as presented here (Lemmas 1 and 2 and Theorems 3, 4, 6, and 8) is the work of Nevanlinna 1920 [1-B-989]. Gronwall 1916 [1-B-453 and 454] announced that he had proved Theorems 3 and 8, but the proof that he sketches seems to be in error (see Goluzin 1951 [1-B-415, p. 351]). Gronwall's proof for convex functions (see Chapter 8), which follows the same pattern, is correct.

Theorem 9. If $f(z)$ is in S, then for any z in E

$$\frac{1-r}{1+r} \leq \left|\frac{zf'(z)}{f(z)}\right| \leq \frac{1+r}{1-r}. \tag{29}$$

Both sides of (29) are sharp. Equality occurs for $f(z) = z/(1-z)^2$.

Proof. We consider the function $g(z)$ defined by equation (6). Since $g(z)$ is in S we may apply Theorem 8 at the point $z = -\lambda$. We obtain

$$\frac{|\lambda|}{(1+|\lambda|)^2} \leq |g(-\lambda)| = \left|\frac{-f(\lambda)}{f'(\lambda)(1-|\lambda|^2)}\right| \leq \frac{|\lambda|}{(1-|\lambda|)^2}. \tag{30}$$

Since λ is an arbitrary point in E, we may replace λ by z in (30). After a few simple steps, the inequality (30) will yield (29). If $f(z)$ is the Koebe function, then $zf'(z)/f(z) = (1+z)/(1-z)$. ■

We can use the upper bound given in Theorem 9 to refine the results obtained so far. We first observe that (29) yields

$$\left|\frac{f'(z)}{f(z)}\right| \leq \frac{1}{r}\frac{1+r}{1-r} = \frac{\psi'(r)}{\psi(r)}, \tag{31}$$

if $\psi(r) \equiv r/(1-r)^2$. We now suppress the middle term in (31) and suppose that $\psi(r)$ has some reasonable properties as set forth in

Theorem 10. Biernacki[1] Suppose that for $0 \leq r < R$, $\psi(r)$ is a real positive differentiable function of r, and that in this interval $f(z)$ satisfies the inequality

[1] Krzyz 1955 [1-B-757] states that Theorem 10 and the method used in Theorems 11 and 12 are due to Biernacki. However no reference is given.

(32) $$\left|\frac{f'(z)}{f(z)}\right| \leq \frac{d}{dr} \ln \psi(r) = \frac{\psi'(r)}{\psi(r)}, \qquad z = re^{i\theta}.$$

Then each of the quantities

(33) $$Q_1(r) \equiv \frac{|f(re^{i\theta})|}{\psi(r)} \qquad \text{and} \qquad Q_2(r) \equiv \frac{M(r, f)}{\psi(r)}$$

is a nonincreasing function of r in $(0, R)$. Further, $\lim_{r \to R^-} Q_j(r)$ exists $(j = 1, 2)$.

We observe that in $Q_1(r)$, θ is fixed. We recall from Problem 3.7 that

(34) $$M(r, f) \equiv \max_{|z|=r} |f(z)|.$$

Proof. Consider first $Q_1(r)$. We have (with θ fixed)

(35) $$\frac{\partial}{\partial r} \ln |f(re^{i\theta})| = \frac{\frac{\partial}{\partial r}|f(re^{i\theta})|}{|f(re^{i\theta})|} \leq \frac{|f'(re^{i\theta})|}{|f(re^{i\theta})|} \leq \frac{d}{dr} \ln \psi(r).$$

Hence $(\ln Q_1(r))' \leq 0$ and $Q_1(r)$ is nonincreasing.

Suppose that $r_1 < r_2$. Select θ so that $|f(r_2 e^{i\theta})| = M(r_2, f)$. Then

(36) $$Q_2(r_2) \equiv \frac{M(r_2, f)}{\psi(r_2)} = \frac{|f(r_2 e^{i\theta})|}{\psi(r_2)} \leq \frac{|f(r_1 e^{i\theta})|}{\psi(r_1)} \leq \frac{M(r_1, f)}{\psi(r_1)} \equiv Q_2(r_1).$$

It is now trival that $\lim Q_j(r)$ exists as $r \to R^-$. ∎

Theorem 11. Hayman 1953 [1-B-511] and 1955 [1-B-512]. If $f(z)$ is in S, then

(37) $$\lim_{r \to 1^-} M(r, f) \frac{(1-r)^2}{r} = \alpha \leq 1,$$

and the quantity on the left is nonincreasing. Further, $\alpha = 1$ if and only if $f(z)$ is some rotation of the Koebe function.

This theorem is often called Hayman's Regularity Theorem.

Proof. We apply Theorems 9 and 10 with $\psi(r) = r/(1-r)^2$. ∎

Theorem 12. Krzyz [1-B-757]. If $f(z)$ is in S, then

$$\lim_{r\to 1^-} M(r, f')(1-r)^3 = \beta \le 2. \tag{38}$$

Proof. From Theorem 6, we have

$$\left|\frac{f''(z)}{f'(z)}\right| \le \frac{4+2r}{1-r^2} = \frac{d}{dr} \ln \frac{1+r}{(1-r)^3} \tag{39}$$

for $0 \le r < 1$. Then by Theorem 10, applied to $f'(z)$, we find that $M(r, f')(1-r)^3/(1+r)$ is nonincreasing in $[0, 1)$ and hence has a limit as $r \to 1^-$. Further, if γ is the limit, then $\gamma \le 1$ with equality if and only if $f(z)$ is a rotation of $k(z)$. If we drop the factor $1/(1+r)$, then the limit β exists and clearly $\beta = 2\gamma \le 2$. ■

Further, Krzyz [1-B-757] proved that if $f(z)$ is in S, then $\beta = 2\alpha$.

Any number in the interval [0, 1] can be the limit described in Theorem 11 if $f(z)$ is selected properly. Indeed (see Problem 2.32), let

$$f(z) \equiv \frac{z - Az^2}{(1-z)^2}, \qquad 0 \le A \le 1. \tag{40}$$

Then $f(z)$ is in S, $M(r, f) = r(1-Ar)/(1-r)^2$, and

$$\lim_{r\to 1^-} M(r, f)(1-r)^2 = 1 - A = \alpha, \tag{41}$$

if $A = 1 - \alpha$. Further,

$$\lim_{r\to 1^-} M(r, f')(1-r)^3 = 2(1-A) = 2\alpha. \tag{42}$$

As a companion for Theorems 11 and 12, Dinghas 1959 [1-B-247] proved that for the minimum of $|f(z)|$: If $f(z)$ is in S, then the two functions

$$m(r, f)\frac{(1+r)^2}{r} \qquad \text{and} \qquad m(r, f')\frac{(1+r)^3}{1-r} \tag{43}$$

are nondecreasing for r in $[0, 1)$.

With much more labor Hayman proved

Theorem 13. **PWO.** Hayman [1-B-512]. If $f(z)$ is in S, then

(44)
$$\lim_{n\to\infty} \frac{|a_n|}{n} = \alpha$$

where α is the number determined in Theorem 11.

Further, Hayman extended the results in Theorem 11 and 13 to mean p-valent functions. For details see Hayman 1958 [1-B-515, pp. 94-116].

4. A ROBERTSON CONJECTURE

We know from Fekete and Szegö that if

(45)
$$g(z) = z + \sum_{k=2}^{\infty} b_{2k-1} z^{2k-1}$$

is an odd univalent function, then the assertion $|b_{2k-1}| \leq 1$ is false (see the answer to Problem 2.23). Nevertheless, when $|b_5| > 1$, it turns out that $|b_3| < 1$ and so there is hope that some weaker inequality combining b_3 and b_5 might still be true. Following these ideas Robertson 1936 [1-B-1141] proposed what we will call

Robertson's conjecture for $S^{(2)}$. If $f(z)$ given by (45) is in $S^{(2)}$, then for all n

(46)
$$1 + \sum_{k=2}^{n} |b_{2k-1}|^2 \leq n.$$

Since $|b_3| \leq 1$ by Theorem 4.3, the conjecture is true for $n = 2$. In [1-B-1141] Robertson proved that $1 + |b_3|^2 + |b_5|^2 \leq 3$, so the conjecture is true when $n = 3$. After a lapse of thirty-four years S. Friedland 1970 [2-B-405] proved the conjecture for $n = 4$. Interest in this conjecture is increased by

Theorem 14. If Robertson's conjecture is true for any fixed n, then the conjecture $|a_n| \leq n$ for every $f(z)$ in S is also true for the same n.

Proof. Robertson [1-B-1141]. Suppose that $f(z)$ is in S and as usual $f(z) = z + \Sigma_2^\infty a_k z^k$. Let $g(z) = \sqrt{f(z^2)} = z + \Sigma_2^\infty b_{2k-1} z^{2k-1}$ be the associated odd univalent function. Then

(47) $$f(z^2) = g^2(z) = z^2 + \sum_{n=2}^{\infty} a_n z^{2n} = \left(z + \sum_{k=2}^{\infty} b_{2k-1} z^{2k-1}\right)^2.$$

Equating coefficients of the same powers of z we get

(48) $$a_n = \sum_{k=1}^{n} b_{2k-1} b_{2n-2k+1}, \qquad b_1 = 1.$$

Then the CBS inequality gives

(49) $$|a_n| \le \left(\sum_{k=1}^{n} |b_{2k-1}|^2\right)^{1/2} \left(\sum_{k=1}^{n} |b_{2n-2k+1}|^2\right)^{1/2} = \sum_{k=1}^{n} |b_{2k-1}|^2.$$

Thus the inequality (46) implies $|a_n| \le n$ for every $f(z)$ in S. ■

PROBLEMS

1. Prove Theorem 2. Hint: First assume that $a_2 \ge 0$. Then deduce the general case by a rotation.
2. Let $f(z)$ be a normalized odd univalent function. Find the radius of the largest disk with center at the origin that is always covered by $f(E)$.
3. Let $f(z)$ be in S and suppose that $f(z)$ omits two values γ and δ. Prove that

   (50) $$\left|\frac{1}{\gamma} - \frac{1}{\delta}\right| \le 4.$$

4. Suppose that in Problem 3, $\gamma = r_1 e^{i\alpha}$ and $\delta = -r_2 e^{i\alpha}$ (the omitted points are collinear and separated by the origin). Prove that $\max(r_1, r_2) \ge 1/2$.
5. Give some justification for the inequalities in (27).
6. Extend Theorems 3, 6, 8, and 9 by proving that in each theorem the equality sign occurs if and only if $f(z)$ is some rotation of the Koebe function.
7. Show that for the Koebe function $|\arg f'(z)| \le 2\pi$ for z in E.
8. Show that for the function discussed in Problem 2.5 (with $\alpha = \pi/4$), $\arg f'(z)$ is unbounded in E. What can you say about $\arg(F(z)/z)$, where $F(z) = f(z) - 1$?
9. Use Problem 2.32 to prove that

$$G(z) \equiv \int_0^z \frac{1+\zeta e^{i\alpha}}{(1-\zeta e^{-i\alpha})^3}\, d\zeta = \frac{z+(i \sin \alpha)z^2}{(1-ze^{-i\alpha})^2}$$

is in S. Use $G(z)$ to prove that the first inequality in Theorem 5 is sharp. Use Problem 8 to prove that $G(z)$ cannot supply the maximum for $|\arg f'(z)|$ for every r in $(0, 1)$. Goluzin [1-B-386, p. 131].

10. Find sharp upper and lower bounds for $|f(z)|$ if $f(z)$ is in $S^{(k)}$.

11. Find sharp upper and lower bounds for $|zf'(z)/f(z)|$ if $f(z)$ is in $S^{(k)}$. Hint: See Theorem 9.

★ **12.** Find sharp upper and lower bounds for $|f'(z)|$ if $f(z)$ is in $S^{(k)}$.

★ **13.** Show that Theorem 2 is sharp by proving that each point on the boundary of the covered disk is an omitted value for some function in S with the prescribed a_2. Hint: It is sufficient to consider the case $a_2 > 0$.

14. Prove that if $f(z)$ is in S, then

$$\textbf{(51)} \qquad |f''(z)| \leq \frac{2(2+r)}{(1-r)^4} \quad \text{and} \quad |f'(z)-1| \leq \frac{4r-3r^2+r^3}{(1-r)^3},$$

and these inequalities are sharp. Bieberbach 1919 [1-B-118]. Using the fact that $|a_3| \leq 3$ and more computation, Landau 1926 [1-B-808] proved that if $f(z)$ is in S, then $|f'''(z)| \leq (18+6r)/(1-r)^5$ and this is sharp. Further Landau proved that if we know that $|a_n| \leq n$ for $n = 2, 3, 4, \cdots, N$ for $f(z)$ in S, then for $f(z)$ in S

$$\textbf{(52)} \qquad |f^{(N)}(z)| \leq k^{(N)}(r).$$

Thus (52) is now known to be true for $N = 2, 3, 4, 5$, and 6. See also Marty 1932 [1-B-906].

15. Show that $-\ln(1-z) = \Sigma_{n=1}^{\infty} z^n/n$ is in S.

★ **16.** Suppose that $f(z)$ is in S and that, further, A, the area of $f(E)$, is finite. Prove that under these conditions

$$\textbf{(53)} \qquad |f(z)| \leq \sqrt{\frac{A}{\pi} \ln \frac{1}{1-r^2}}$$

and this inequality is sharp. Hint: Use the CBS inequality. The extremal function depends on r. Gronwall 1914 [1-B-452].

★ **17.** Under the conditions of Problem 16, prove that

$$\textbf{(54)} \qquad |f'(z)| \leq \sqrt{\frac{A}{\pi}} \frac{1}{1-r^2}.$$

Hint: The extremal function depends on r and is not the same as the one for Problem 16. Gronwall [1-B-452].

18. Prove that for any analytic function

$$\frac{\partial}{\partial r}|f(z)| \leq |f'(z)|. \tag{55}$$

This was used in the proof of Theorem 10 (equation (35)).

19. Let $0 \leq A \leq 1$ and let

$$g(z) = \frac{1-Az}{1-z} \tag{56}$$

Prove that $M(r, g) = (1-Ar)/(1-r) = g(r)$.

20. Let $f(z)$ be defined by equation (40). Use the result in Problem 19 to prove that $M(r, f) = f(r)$, and $M(r, f') = f'(r)$.

21. Let $b(z) \equiv (1-A)z/(1-Az)$ where $0 < A < 1$. Prove that for this function $|b(z)| < 1$ for z in E. It follows trivially that if z_1 and z_2 are in E, then $b(z_1)b(z_2) \neq 1$.

22. Let $f(z)$ be defined by equation (40). If $A = 0$ or $A = 1$, the function is trivially in S. Prove that if $0 < A < 1$ and $f(z_1) = f(z_2)$, then $b(z_1)b(z_2) = 1$ where $b(z)$ is the function considered in Problem 21. Hence $f(z)$ is in S. Of course, this same result is a special case of Problem 2.32, but we now have a different proof.

★ **23.** Suppose that $f(z)$ is a normalized function, regular in E but not necessarily univalent in E. Suppose that the complement of $f(E)$ contains the sector that has its vertex at $w_0 = -R < 0$, vertex angle $\alpha < 2\pi$, and is symmetric with respect to the real axis. Thus $f(z)$ omits in E all w for which $\pi - \alpha/2 \leq \arg(w - w_0) \leq \pi + \alpha/2$. Prove that $R \geq 1/(4 - 2\alpha/\pi)$ and that this result is sharp for each α in $[0, 2\pi)$. Observe that when $\alpha = 0$ we obtain $R \geq 1/4$. See Szilárd 1969 [2-B-1413].

24. Find sharp upper and lower bounds for $|f(z)|$, if $f(z)$ is in $SB_1(M)$.

25. Let ES be the set of all functions $f(z)$ in S for which every derivative $f^{(k)}(z)$ is also univalent in E. Prove that $F(z) \equiv (e^{\pi z} - 1)/\pi$ is in ES. Find a_2 for $F(z)$. S. M. Shah and Trimble 1969 [2-B-1306] for Problems 25 through 28.

26. Let $A_2 = \sup |a_2|$ for all functions in ES. Prove that for $k = 0, 1, 2, \ldots$, we have $|a_{k+2}| \leq 2|a_{k+1}|A_2/(k+2)$.

27. Prove that if $f(z)$ is in ES, then it is an entire function. Hint: Find a bound for $|a_k|$ and $|f(z)|$.

★ **28.** Prove that $A_2 < 1.7208$. Hint: Use the bound on $|f(z)|$ in E and Problem 4.13. Note that Problem 25 gives $\pi/2 \approx 1.570796 < A_2 < 1.7208$. The determination of the true value of A_2 is an open problem. The attractive conjecture that $A_2 = \pi/2$ was destroyed by Lachance 1980 [3-B-495], who proved that $1.5910 < A_2$. Sathaye and S. M. Shah [1] pushed this lower bound up to 1.593905. In a series of papers S. M. Shah and Trimble proved many interesting theorems about the set *ES*. For a survey and further references, see S. M. Shah and Trimble 1978 [3-B-807]. See also G. A. Read 1969 [2-B-1192] and B. S. Lee 1972 [2-B-817]. For an interesting extension of these ideas to linear invariant families of functions, see Campbell 1971 [2-B-206].

CHAPTER 7

FUNCTIONS WITH POSITIVE REAL PART

1. ELEMENTARY PROPERTIES OF THE SET *P*

Suppose that $g(z)$ is regular in E and that $g(E)$ is contained in some open half-plane. We can ask (and answer) many questions concerning functions of this type. As in the case of the study of univalent functions it is convenient to introduce some normalization. Here we replace the arbitrary half-plane by the half-plane H^+: Re $w > 0$, and we require that $f(0) = 1$, where $f(z)$ is the normalized function.

To achieve this normalization let us suppose that $g(E)$ lies in the open half-plane bounded by the line L: $w = A + e^{i\alpha}t$. Here α and t are real, the line L is oriented by the agreement that L is described as t increases from $-\infty$ to ∞, and $g(E)$ lies to the right of L. Under these conditions, we set $h(z) = e^{-i\alpha}(g(z) - A)$ and observe that $h(E)$ lies in H^+ (hereafter called the right half-plane). Let $h(0) = C + iD$. Here $C > 0$, because $h(E) \subset H^+$. Consequently, if we set $f(z) = (h(z) - iD)/C$, we have our normalized function: $f(E) \subset H^+$ and $f(0) = 1$.

Definition 1. The set P is the set of all functions of the form

$$(1) \quad f(z) = 1 + p_1 z + p_2 z^2 + \cdots + p_n z^n + \cdots = 1 + \sum_{n=1}^{\infty} p_n z^n$$

that are regular in E, and such that for z in E,

$$(2) \qquad \operatorname{Re}(f(z)) > 0.$$

Any function in P is called a function with *positive real part* in E.

It should be noted that $f(z) = 1 + z^n$ is in P for any integer $n \geq 0$, but if $n \geq 2$, this function is *not* univalent. It is clear that with sufficient labor, any theorem about the class P can be transformed into a suitable theorem about a function $g(z)$ that carries E into some prescribed half-plane.

Just as the Koebe function plays a central role in the class S, the Möbius function

$$(3) \qquad L_0(z) \equiv \frac{1+z}{1-z} = 1 + 2z + 2z^2 + \cdots = 1 + 2\sum_{n=1}^{\infty} z^n$$

plays a central role in the class P. This function is in the class P, it is regular and univalent in E, and it maps E onto the half-plane H^+. There is one notable difference in the character of $L_0(z)$ and $k(z)$. In many extremal problems for the class S, the Koebe function is the unique solution (apart from a rotation). In contrast, the function $L_0(z)$ does maximize $|p_n|$ in the class P, but if $n \geq 2$, there are infinitely many other functions in P for which $p_n = 2$, and no one of these is obtainable from any other by a rotation (see Theorem 3).

We begin by observing that the set P is *convex*. By this we mean that if μ_1 and μ_2 are nonnegative with $\mu_1 + \mu_2 = 1$ and $f_1(z), f_2(z)$ are in P, then

$$(4) \qquad f(z) \equiv \mu_1 f_1(z) + \mu_2 f_2(z)$$

is also in P. It is a trivial matter to pass from equation (4) to a finite sum, and then to an infinite sum

$$(5) \qquad f(z) = \sum_{k=1}^{\infty} \mu_k f_k(z),$$

if we assume that $\mu_k \geq 0$ for each k, and $\Sigma_{k=1}^{\infty} \mu_k = 1$.

We can go further and replace (5) by a Stieltjes integral if the functions $f_k(z)$ are selected properly. It is clear that if $f(z)$ is in P, then for each real ϕ, $f(e^{-i\phi}z)$ is in P. We apply this remark to the function $L_0(z) \equiv (1+z)/(1-z)$ and extend (5) to a Stieltjes integral in which the positive weight μ_k is replaced by $d\mu(\phi)$. This gives

Theorem 1. PLE. Let $\mu(\phi)$ be a nondecreasing function in the interval $0 \leq t \leq 2\pi$, with

$$\int_0^{2\pi} d\mu(\phi) = 2\pi; \tag{6}$$

then the function

$$f(z) \equiv \frac{1}{2\pi}\int_0^{2\pi} \frac{1+ze^{-i\phi}}{1-ze^{-i\phi}}\, d\mu(\phi) \equiv \frac{1}{2\pi}\int_0^{2\pi} L_0(e^{-i\phi}z)\, d\mu(\phi) \tag{7}$$

is in P.

The remarkable fact, proved by Herglotz, 1911 [1-B-526], is that the converse of this theorem is also true: for each $f(z)$ in P there is an associated nondecreasing function $\mu(\phi)$ for which (6) and (7) hold. Since the proof is somewhat complicated, we postpone it to the end of the chapter.

Some operations that carry functions in P into functions in P are given in

Theorem 2. PLE. Suppose that $f(z)$, $f_1(z)$, and $f_2(z)$ are in P. Then, subject to the conditions noted, the function $g(z)$ is also in P, where:

$$g(z) = f(e^{i\alpha}z), \qquad \alpha \text{ real}, \tag{8}$$

$$g(z) = [f(z)]^t, \text{ or } g(z) = f(tz), \qquad -1 \leq t \leq 1, \tag{9}$$

$$g(z) = 1/f(z), \tag{10}$$

$$g(z) = [f_1(z)]^{t_1}[f_2(z)]^{t_2}, \qquad 0 \leq t_1, t_2, \quad t_1 + t_2 \leq 1, \tag{11}$$

$$g(z) = \frac{1}{a}\left[f\left(\frac{z+\lambda}{1+\bar{\lambda}z}\right) - bi\right], \qquad \text{if } f(\lambda) = a + bi, \quad \lambda \in E, \tag{12}$$

$$g(z) = \frac{f(z)+ib}{1+ibf(z)}, \qquad b \text{ real} \tag{13}$$

To prove that $g(z)$, defined by (13), is in P whenever $f(z)$ is in P, we observe that the linear transformation

$$W = \frac{w + ib}{1 + ibw}$$

(with b real) maps H^+ onto H^+ and takes $w = 1$ into $W = 1$. Then we apply this transformation to $w = f(z)$.

2. THE COEFFICIENT BOUNDS

These are given in

Theorem 3. Carathéodory 1907 [1-B-185]. Let $N \geq 1$ be a fixed integer. If $f(z)$, given by (1), is in P, then

$$|p_N| \leq 2. \tag{14}$$

This inequality is sharp. If $\eta = e^{2\pi i/N}$ and

$$F(z) \equiv \sum_{k=1}^{N} \mu_k \frac{1 + \eta^k z}{1 - \eta^k z} \equiv 1 + \sum_{n=1}^{\infty} P_n z^n, \tag{15}$$

where $\mu_k \geq 0$, for $k = 1, 2, \ldots, N$, and

$$\sum_{k=1}^{N} \mu_k = 1, \tag{16}$$

then $F(z)$ is in P and $P_N = 2$.

Proof. We can always select a real α in $f(e^{i\alpha}z)$ so that $p_N e^{i\alpha N} \geq 0$. Hence we may suppose that $p_N \geq 0$. Consider the integral

$$I(r) \equiv \int_0^{2\pi} (1 - \cos N\theta) f(re^{i\theta})\, d\theta, \qquad 0 \leq r < 1. \tag{17}$$

On the one hand, $\operatorname{Re}(f(re^{i\theta})) > 0$ and hence $\operatorname{Re}(I(r)) > 0$. On the other hand, by the orthogonality of the trigonometric functions

$$I(r) = \int_0^{2\pi} (1 - \cos N\theta)\left(1 + \sum_{k=1}^{\infty} p_k r^k (\cos k\theta + i \sin k\theta)\right) d\theta = 2\pi - \pi r^N p_N.$$

Thus $\pi(2 - r^N p_N) = \operatorname{Re}(I(r)) > 0$. Consequently $r^N p_N < 2$. Since we may take r as close to 1 as we please, we find that $0 \le p_N \le 2$. This gives the inequality (14).

To prove that $F(z)$ is in P, we need only to observe that (15) is a special case of the Stieltjes integral representation (7). Finally, it is easy to see that the Nth coefficient of $F(z)$ is given by

$$(18) \qquad p_N = 2\sum_{k=1}^{N} \mu_k (\eta^k)^N = 2\sum_{k=1}^{N} \mu_k (\eta^N)^k = 2\sum_{k=1}^{N} \mu_k = 2. \qquad \blacksquare$$

Two different selections of the ordered N-tuple $(\mu_1, \mu_2, \ldots, \mu_N)$ lead to functions $F_1(z)$ and $F_2(z)$ that are different. Indeed the two functions may have the same poles, but there will be at least one pole where the two functions have different residues. Thus the set of functions defined by (15) depend upon $N-1$ independent real parameters.[1] It can be shown that if equality occurs in (14), then $f(z)$ must have the form $F(e^{i\alpha}z)$ for some function of the type defined by equations (15) and (16) (see Problem 4).

If we write the integrand in (7) as an infinite series and integrate termwise (with $|z| < 1$), we find that the coefficients in (1) are given by

$$(19) \qquad p_n = \frac{1}{\pi}\int_0^{2\pi} e^{-in\phi}\, d\mu(\phi).$$

Thus equations (6), (7), and (19) will lead to a new proof that $|p_N| \le 2$. A third proof will be given after we have proved Theorem 11.

3. THE TECHNIQUE OF DOMINANT POWER SERIES

The sharp bounds for the coeficients make it easy to prove

Theorem 4. If $f(z)$ is in P and $z = re^{i\theta}$, then

$$(20) \qquad \frac{1-r}{1+r} \le |f(z)| \le \frac{1+r}{1-r}$$

and

[1] In ancient times this was referred to as an $(N-1)$-infinity of extremal functions.

(21) $$|f'(z)| \le \frac{2}{(1-r)^2},$$

and these inequalities are sharp. Equality occurs for suitable z if and only if $f(z) = L_0(e^{i\alpha}z)$.

As in Theorem 3, there are several ways of proving this theorem. Here we are interested in using a new and very simple technique, the technique of dominant power series.

Definition 2. Let $f(z)$ and $F(z)$ be given by the power series

(22) $$f(z) = \sum_{n=0}^{\infty} a_n z^n$$

and

(23) $$F(z) = \sum_{n=0}^{\infty} A_n z^n,$$

convergent in some disk E_R: $|z| < R$, $R > 0$. We say that $f(z)$ is *dominated* by $F(z)$ (or $F(z)$ *dominates* $f(z)$), and we write $f(z) << F(z)$ if for each integer $n \ge 0$,

(24) $$|a_n| \le A_n.$$

A number of simple statements about dominance are collected and presented in

Theorem 5. PLE. Suppose that $f(z) << F(z)$. Then

(25) $$A_n \ge 0, \qquad n = 0, 1, 2, \ldots,$$

(26) $$|f(z)| \le F(r), \qquad 0 \le |z| \equiv r < R,$$

(27) $$f'(z) << F'(z),$$

(28) $$\int_0^z f(\zeta)\, d\zeta << \int_0^z F(\zeta)\, d\zeta,$$

(29) $$e^{f(z)} << e^{F(z)},$$

and if $|F(z)| < 1$ in some disk E_ρ, then in that disk

(30) $$-\ln(1 - f(z)) << -\ln(1 - F(z)).$$

Further, if $|b| \le B$ and $k \ge 0$ is any integer, then

$$bz^k f(z) << Bz^k F(z) \tag{31}$$

and

$$[f(z)]^k << [F(z)]^k. \tag{32}$$

Finally, if $f(z) << F(z)$ and $g(z) << G(z)$, then

$$f(z) + g(z) << F(z) + G(z) \tag{33}$$

and

$$f(z)g(z) << F(z)G(z). \tag{34}$$

Proof of Theorem 4. Since $|p_n| \le 2$, it is clear that

$$f(z) << L_0(z) \equiv \frac{1+z}{1-z}. \tag{35}$$

Then (26) gives the right side of (20) and (27) with (26) gives the right side of (21). For the lower bound in (20), we apply the right side of (20) to the function $1/f(z)$, which is in P whenever $f(z)$ is in P. It follows that $1/|f(z)| \le (1+r)/(1-r)$ and hence $|f(z)| \ge (1-r)/(1+r)$. From Problem 5, equality can occur if and only if $f(z) = L_0(e^{i\alpha}z)$ and z is selected properly ($z = \pm e^{-i\alpha}r$). ■

Since $f'(z)$ may be zero for a function in P, it is clear that no positive lower bound can be given in (21) unless some additional hypothesis is made about $f(z)$.

By differentiating $f(z)$ k times and using (27) each time, we obtain

Theorem 6. PLE. If $f(z)$ is in P and $k \ge 0$, then

$$|f^{(k)}(z)| \le \frac{d^k}{dz^k}\left(\frac{1+z}{1-z}\right)\bigg|_{z=r} = \frac{2(k!)}{(1-r)^{k+1}}. \tag{36}$$

In particular, if $k = 2$ we have

$$|f''(z)| \le \frac{4}{(1-r)^3}. \tag{37}$$

4. SUBORDINATION AND THE LINDELÖF PRINCIPLE

Using this principle we will soon see that the ring domain specified for $f(z)$ by the inequality (20) can be reduced to the disk shown in Figure 1.

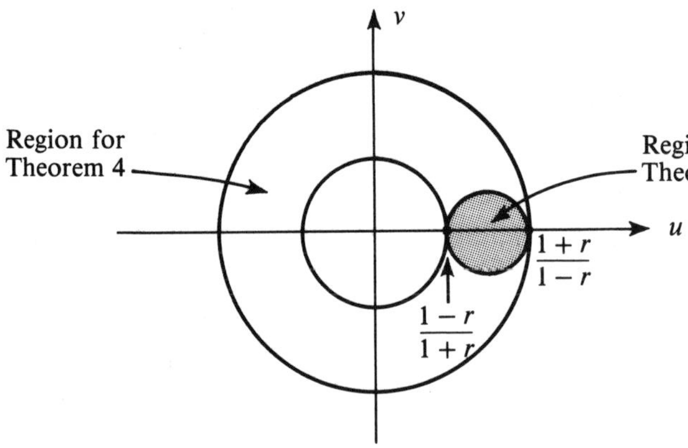

FIGURE 1.

Theorem 7. If $f(z)$ is in P, then for each fixed z in E, $f(z)$ lies in the closed disk with center at $(1 + r^2)/(1 - r^2)$ and radius $2r/(1 - r^2)$. Further, if $z \neq 0$, then $f(z)$ is a boundary point of this disk if and only if $f(z) = L_0(e^{i\alpha}z)$ for some real α.

Thus, if $f(z)$ is in P, then it satisfies the sharp inequality

$$\left| f(z) - \frac{1 + r^2}{1 - r^2} \right| \leq \frac{2r}{1 - r^2}.$$

The diameter end points of the disk that contains $f(z)$ are $(1 - r)/(1 + r)$ and $(1 + r)/(1 - r)$.

To prove this theorem we use the Lindelöf principle (see Theorem 10), which is really a natural extension of Schwarz's Lemma.

Theorem 8. Schwarz's Lemma. Let $b(z)$ be in B_0. Thus $b(z)$ is regular in E, $b(0) = 0$, and $|b(z)| \leq 1$ for z in E. Then for each r with $0 < r < 1$,

$$|b(re^{i\theta})| \leq r. \tag{38}$$

If equality occurs in (38) for one point $z_0 = re^{i\theta}$ with $0 < r < 1$, then $b(z) \equiv e^{i\alpha}z$ for some real α. Finally, we have

$$|b_1| = |b'(0)| \leq 1, \tag{39}$$

and $|b_1| = 1$ if and only if $b(z) = B(z) \equiv e^{i\alpha}z$.

Theorem 8 tell us that the univalent function $B(z) = e^{i\alpha}z$ is in some sense a maximal function among all bounded functions with $b(0) = 0$, or the bounded functions $b(z)$ are "subordinate" to the univalent function $B(z)$. We make this concept of subordination precise, and we extend it to an arbitrary function in

Definition 3. Let $F(z) = a_0 + a_1 z + \cdots$ be regular and univalent in E and suppose that $F(E) = D$. If $f(z)$ is regular in E, $f(0) = F(0)$, and $f(E) \subset D$, then we say that $f(z)$ is *subordinate* to $F(z)$ in E, and we write

$$f(z) \prec F(z). \tag{40}$$

We also say that $F(z)$ is *superordinate* to $f(z)$ in E.

We observe that in this definition $F(z)$ is univalent in E, but nothing is assumed about the valence of $f(z)$. Both $F(z)$ and $f(z)$ carry $z = 0$ into the same point, and $f(z)$ carries E onto some (possibly multi-sheeted) surface whose projection onto the plane is contained in D. For example, under the conditions on $b(z)$ in Schwarz's Lemma, $b(z) \prec B(z) = e^{i\alpha}z$.

Suppose now that $f(z) \prec F(z)$ and $F(E) = D$. Then the inverse F^{-1} is regular in D and maps D onto E with $F^{-1}(a_0) = 0$. Hence the composite function $b(z) \equiv F^{-1}(f(z))$ is regular in E, and maps E into E. Further $b(0) = F^{-1}(f(0)) = F^{-1}(a_0) = 0$. Thus $b(z)$ satisfies the conditions of Schwarz's Lemma, and $f(z) = F(b(z))$. We have proved the "only if" part of

Theorem 9. Let $f(z)$ and $F(z)$ be regular in E, and suppose that $F(z)$ is univalent in E. Then $f(z) \prec F(z)$ in E if and only if there is a function $b(z)$ that satisfies the conditions of Schwarz's Lemma and

$$f(z) = F(b(z)). \tag{41}$$

The proof of the "if" part of this theorem is trivial. ∎

We note that in the definition of subordination we assume that $F(z)$ is univalent in E. The concept of subordination can be extended to the case where $F(z)$ is not univalent, and the simplest way to do this is to use equation (41) as the definition of $f(z) \prec F(z)$. Thus, $f(z) \prec F(z)$ if and only if there is a $b(z)$ in B_0 such that $f(z) = F(b(z))$.

As an example, if n is a positive integer, then $z^n \prec z$ in E. If we do not demand that $F(z)$ is univalent, then $z^{2n} \prec z^2$ in E, but z^{2n+1} is *not* subordinate to z^2 in E.

Theorem 10. The Lindelöf Principle. 1908 [1-B-860]. Suppose that $f(z) \prec F(z)$ in E. Then for each r in $[0, 1]$

$$f(E_r) \subset F(E_r). \tag{42}$$

Further, if $f(re^{i\theta})$ is on the boundary of $F(E_r)$ for one point $z_0 = re^{i\theta_0}$, with $0 < r < 1$, then there is a real α such that $f(z) \equiv F(e^{i\alpha}z)$, and $f(re^{i\theta})$ is on the boundary of $F(E_r)$ for every point $z = re^{i\theta}$ in E.

This theorem is illustrated in Figure 2. Thus, if $f(E) \subset F(E) = D$, then $f(E_r)$ is contained in the shaded region that represents $F(E_r)$.

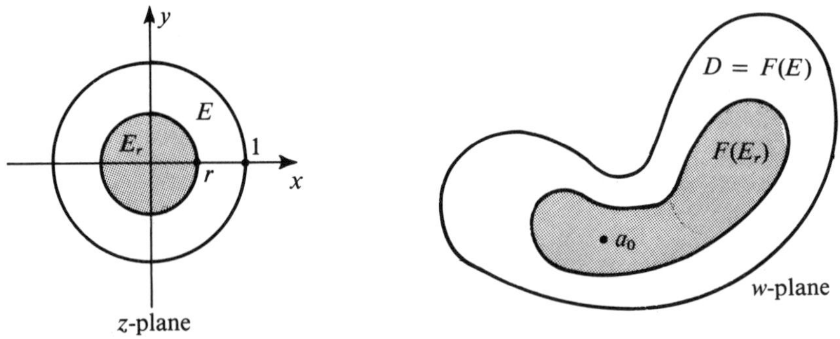

FIGURE 2.

Proof of Theorem 10. If $f(z) \prec F(z)$, then, by Theorem 9, there is a $b(z)$ such that $f(z) = F(b(z))$ and $b(z)$ satisfies the conditions of Schwarz's Lemma (Theorem 8). Then $f(E_r) = F(b(E_r))$. But $b(E_r) \subset E_r$. Hence $f(E_r) = F(b(E_r)) \subset F(E_r)$, and this is (42).

If there is a z_0 such that $0 < r = |z_0| < 1$ and $f(z_0)$ is a boundary point of $F(E_r)$, then $|b(z_0)| = r$. Hence $b(z) = e^{i\alpha}z$ for some real α, and $f(z) = F(e^{i\alpha}z)$. ∎

Proof of Theorem 7. This theorem is really an immediate consequence of Theorem 10. Indeed, if $f(z)$ is in P, then by the definition of the class P, $f(z) \prec L_0(z) = (1+z)/(1-z)$. Hence $f(E_r) \subset L_0(E_r)$. Now the closure of $L_0(E_r)$ is just the closed disk described in Theorem 7. Hence $f(re^{i\theta})$ must lie in this disk. If $f(re^{i\theta})$ is on the boundary of $L_0(E_r)$, then, by Theorem 10, $f(z) = L_0(e^{i\alpha}z)$ for some real α. ■

The concepts used in Theorems 8 and 9 can be used to give an alternate proof that $|p_N| \leq 2$ for $f(z)$ in P. We first give a slight extension of Schwarz's Lemma.

Theorem 11. PLE. Let $N \geq 1$ be a fixed integer and assume that

$$b(z) = b_N z^N + b_{N+1} z^{N+1} + \cdots = \sum_{n=N}^{\infty} b_n z^n \tag{43}$$

satisfies the conditions of Schwarz's Lemma. If $0 < r < 1$, then

$$|b(re^{i\theta})| \leq r^N. \tag{44}$$

If equality occurs in (44) for one point $z_0 = re^{i\theta}$ with $0 < r < 1$, then $b(z) = e^{i\alpha}z^N$ and the equal sign holds in (44) for all z in E. Finally,

$$|b_N| = \frac{1}{N!}|b^{(N)}(0)| \leq 1, \tag{45}$$

and $|b_N| = 1$ if and only if $b(z) = e^{i\alpha}z^N$.

Alternate Proof of Theorem 3. Let $f(z)$ be in P and suppose that $f(z)$ has the Maclaurin series (1). Let $\eta = e^{2\pi i/N}$ and define $g(z)$ by

$$g(z) \equiv \frac{1}{N}\Big(f(z) + f(\eta z) + f(\eta^2 z) + \cdots + f(\eta^{N-1}z)\Big). \tag{46}$$

Then $g(z)$ is in P. Further, by equation (1.17)

$$1 + \eta^k + \eta^{2k} + \cdots + \eta^{(N-1)k} = \begin{cases} 0, & \text{if } k \not\equiv 0 \bmod N, \\ N, & \text{if } k \equiv 0 \bmod N, \end{cases} \tag{47}$$

and hence

$$\begin{aligned} g(z) &= \frac{1}{N}\sum_{j=0}^{N-1}\left(1 + \sum_{k=1}^{\infty} p_k(\eta^j z)^k\right) \\ &= 1 + \frac{1}{N}\sum_{k=1}^{\infty}\left(\sum_{j=0}^{N-1}\eta^{jk}\right)p_k z^k = 1 + p_N z^N + p_{2N}z^{2N} + \cdots. \end{aligned}$$

If L_0^{-1} denotes the inverse of $L_0(z)$, then

$$\text{(48)} \quad b(z) \equiv L_0^{-1}(g(z)) = \frac{g(z)-1}{g(z)+1} = \frac{p_N z^N + p_{2N} z^{2N} + \cdots}{2 + p_N z^N + \cdots}$$
$$= \frac{1}{2} p_N z^N + \cdots$$

satisfies the conditions of Theorem 11. The inequality (45) gives $|p_N| \leq 2$ and equality occurs if and only if $g(z) = L_0(e^{i\alpha} z^N) = (1 + e^{i\alpha} z^N)/(1 - e^{i\alpha} z^N)$ for some real α. ■

5. THE NOSHIRO-WARSCHAWSKI THEOREM

What relation does the class P have to the theory of univalent functions? We will explore some of the relations in the chapters that follow. Here we are content to give in Theorem 13 a beautiful and simple sufficient condition for univalence due independently to Noshiro 1935 [1-B-999] and Warschawski 1935 [1-B-1468, p. 312].

Theorem 12. Alexander 1915 [1-B-52, p. 18]. If $f'(z)$ is in P, then $f(z)$ is univalent in E.

Theorem 12 is merely a special case of

Theorem 13. The Noshiro-Warschawski Theorem. Suppose that for some real α we have

$$\text{(49)} \qquad \operatorname{Re}\,(e^{i\alpha} f'(z)) > 0$$

for all z in a convex domain D. Then $f(z)$ is univalent in D.

Proof. Let z_1 and z_2 be any two distinct points in D. We integrate $f'(z)$ along the line segment L: $z = z_1 + t(z_2 - z_1)$, $0 \leq t \leq 1$. Since D is convex, L lies in D. Now $dz = (z_2 - z_1)\,dt$, hence

$$f(z_2) - f(z_1) = \int_L f'(z)\,dz = \int_0^1 f'(z)(z_2 - z_1)\,dt,$$

$$\text{(50)} \qquad f(z_2) - f(z_1) = (z_2 - z_1)e^{-i\alpha} \int_0^1 e^{i\alpha} f'(z)\,dt.$$

But the integral in (50) is not zero because by (49) it has a positive real part. Hence $f(z_1) \neq f(z_2)$. ■

Example. Prove that $f(z) = -z - 2 \ln (1 - z)$ is in S.

Solution. For this function $f'(z) = (1 + z)/(1 - z) = L_0(z) \in P$. Since E is a convex domain, $f(z)$ is univalent in E.

It can be proved that if we relax the condition in Theorem 12 slightly, then $f(z)$ may have an arbitrarily high valence. Indeed, Goodman 1972 [2-B-479] proved that for each $\epsilon > 0$ and each positive integer p there is a function $f(z)$ for which $|\arg f'(z)| < \dfrac{\pi}{2} + \epsilon$ in E and $f(z)$ assumes some value at least p times in E.

Suppose that D is not a convex domain. Then Theorem 13 again fails. Tims 1951 [1-B-1399] proved that for each nonconvex simply-connected domain D with at least two boundary points, there is a function $f(z)$ regular in D, with $\operatorname{Re} f(z) > 0$ in D, that is not univalent in D.

6. DIVIDED DIFFERENCES AND p-VALENT FUNCTIONS

We now present a beautiful generalization of the Noshiro-Warschawski Theorem to p-valent functions. Some mild preparations are necessary.

Definition 4. A function $f(z)$ meromorphic in a domain D is said to be *p-valent* in D (or *multivalent of order p in D*) if for each w_0 (infinity included) the equation $f(z) = w_0$ has at most p roots in D (where the roots are counted in accordance with their multiplicity) and if there is some w_1 such that the equation $f(z) = w_1$ has exactly p roots in D.

One can avoid multiple roots and the point at infinity by using

Theorem 14. PLE. If $f(z)$ is p-valent in a domain D, then there is a finite w^* such that the equation $f(z) = w^*$ has p simple roots in D.

We will need the expansion of the Vandermonde determinant, namely

(51)
$$\begin{vmatrix} 1 & 1 & \cdots & 1 \\ z_0 & z_1 & \cdots & z_p \\ z_0^2 & z_1^2 & \cdots & z_p^2 \\ \cdot & \cdot & \cdots & \cdot \\ \cdot & \cdot & \cdots & \cdot \\ \cdot & \cdot & \cdots & \cdot \\ z_0^p & z_1^p & \cdots & z_p^p \end{vmatrix} = \prod_{0 \le j < k \le p} (z_k - z_j).$$

We also need the generalization of the difference quotient, that is generally referred to as a *divided difference*. Henceforth, $f(z)$ is a regular funcion (∞ is not permitted as a value for z or $f(z)$). We define the symbol $f[z_0, z_1, \ldots, z_p]$ for $p \geq 1$ by

(52)
$$f[z_0, z_1] = \frac{f(z_1) - f(z_0)}{z_1 - z_0},$$

(53)
$$f[z_0, z_1, z_2] = \frac{f[z_1, z_2] - f[z_0, z_1]}{z_2 - z_0},$$

and inductively for $p > 2$ by

(54)
$$f[z_0, z_1, \ldots, z_p] = \frac{f[z_1, z_2, \ldots, z_p] - f[z_0, z_1, \ldots, z_{p-1}]}{z_p - z_0}.$$

Of course, we are assuming that the z_k are distinct, but suitable conventions can be made using limit values and derivatives when some of the z_k coincide. Using induction one can prove

Theorem 15. PLE. Let $z_0, z_1, \ldots, z_p$ be distinct, and for each k, set

(55)

$$D^*(k, n) \equiv (z_k - z_0)(z_k - z_1) \cdots (z_k - z_{k-1})(z_k - z_{k+1}) \cdots (z_k - z_p) = \prod_{\substack{j=0 \\ j \neq k}}^{p} (z_k - z_j).$$

Then

(56)
$$f[z_0, z_1, \ldots, z_p] = \sum_{k=0}^{p} \frac{f(z_k)}{D^*(k, p)}$$

For these same divided differences we also have

Theorem 16. PLE. If $z_0, z_1, \ldots, z_p$ are distinct, then

$$
(57) \qquad f[z_0, z_1, \ldots, z_p] = \frac{\begin{vmatrix} 1 & 1 & \cdots & 1 \\ z_0 & z_1 & \cdots & z_p \\ z_0^2 & z_1^2 & \cdots & z_p^2 \\ \cdot & \cdot & \cdots & \cdot \\ \cdot & \cdot & \cdots & \cdot \\ \cdot & \cdot & \cdots & \cdot \\ z_0^{p-1} & z_1^{p-1} & \cdots & z_p^{p-1} \\ f(z_0) & f(z_1) & \cdots & f(z_p) \end{vmatrix}}{\begin{vmatrix} 1 & 1 & \cdots & 1 \\ z_0 & z_1 & \cdots & z_p \\ z_0^2 & z_1^2 & \cdots & z_p^2 \\ \cdot & \cdot & \cdots & \cdot \\ \cdot & \cdot & \cdots & \cdot \\ \cdot & \cdot & \cdots & \cdot \\ z_0^{p} & z_1^{p} & \cdots & z_p^{p} \end{vmatrix}}
$$

To prove Theorem 16 expand the numerator in (57) by minors of the last row, use the identity (51) for the denominator, and show that the quotient is identical with the right side of (56).

An interchange of columns in both the numerator and denominator of (57) leaves the ratio unchanged. Hence we have

Theorem 17. If $z_{k_0}, z_{k_1}, z_{k_2}, \ldots, z_{k_p}$ is any permutation of $z_0, z_1, z_2, \ldots, z_p$, then

$$
(58) \qquad f[z_{k_0}, z_{k_1}, z_{k_2}, \ldots, z_{k_p}] = f[z_0, z_1, z_2, \ldots, z_p].
$$

The importance of Theorem 16 for p-valent functions lies in

Theorem 18. PLE. If the numerator in (57) is not zero for all sets $\{z_0, z_1, \ldots, z_p\}$ of distinct points in a domain D, then $f(z)$ has valence $q \leq p$ in D.

The converse is false. For example, the function $f(z) \equiv z^2$ is at most 2-valent in any domain, but if $p \geq 3$ the numerator in equation (57) vanishes identically when $f(z) \equiv z^2$.

Theorem 19. PLE. Let $f(z)$ be regular in a domain D, and let z_0, $z_1, \ldots, z_p$ be distinct points in D. Then

(59)

$$f[z_0, z_1, \ldots, z_p] = \int_0^1 \int_0^{t_1} \int_0^{t_2} \cdots \int_0^{t_{p-1}} f^{(p)}(\zeta)\, dt_p dt_{p-1} \cdots dt_2 dt_1,$$

where $\zeta = z_0 + (z_1 - z_0)t_1 + (z_2 - z_1)t_2 + \cdots + (z_p - z_{p-1})t_p$.

The industrious reader may find that he or she must use Theorem 17 in proving Theorem 19 by induction.

The natural generalization of the Noshiro-Warschawski Theorem (Theorem 13) is

Theorem 20. Suppose that $f(z)$ is regular in a convex domain D and that for some real α we have $\operatorname{Re}(e^{i\alpha} f^{(p)}(z)) > 0$ in D. Then $f(z)$ is at most p-valent in D.

Proof. Under these conditions the path of the iterated integral (59) lies in D and hence the iterated integral is never zero. Consequently, the quotient of determinants in (57) is never zero. Then by Theorem 18 the valence of $f(z)$ cannot exceed p. ■

Actually we have proved somewhat more. Let $P_{p-1}(z)$ denote an arbitrary polynomial of degree $\leq p-1$. Since the pth derivative of $P_{p-1}(z)$ vanishes identically, it is clear that $P_{p-1}(z) + f(z)$ satisfies the hypotheses of Theorem 20 whenever $f(z)$ does. Thus if q is the valence of $P_{p-1}(z) + f(z)$, then $q \leq p$ for any polynomial of degree $\leq p-1$. We call such a function *completely p-valent*. Thus any function that satisfies the conditions of Theorem 20 is completely p-valent in the convex domain D.

Theorems 18, 19, and 20 were independently obtained by: Montel 1932 [3-B-621] and 1937 [1-B-941], Ozaki 1935 [1-B-1021], and Kimura 1937 [1-B-668].

Let $K_p(D)$ be the set of all functions that are regular in a domain D and satisfy $f[z_0, z_1, \ldots, z_p] \neq 0$ for all sets of points $z_0, z_1, \ldots, z_p$

in D. Let $\widetilde{P}(E)$ be the set of all functions $f(z) = z + \cdots$ regular in E for which $z^{p-1}f(z)$ is in $K_p(E)$ for $p = 1, 2, \ldots$. For work on these two sets see Kirjackis [1-B-A-65, 66, and 67] and [2-B-700 and 701], Gopengauz [2-B-493 and 494], and Ruscheweyh 1976 [3-B-767].

7. EXTREME POINTS AND THE KREIN-MIL'MAN THEOREM

This important theorem is the complete generalization to a topological vector space of a rather elementary theorem on convex sets in the plane. To make matters simple we first look at the theorem in the plane. As indicated in Figure 3, C is a closed and bounded convex set and L_1, L_2, and L_3 are lines in the plane. The point Q_1 is an interior point of the line

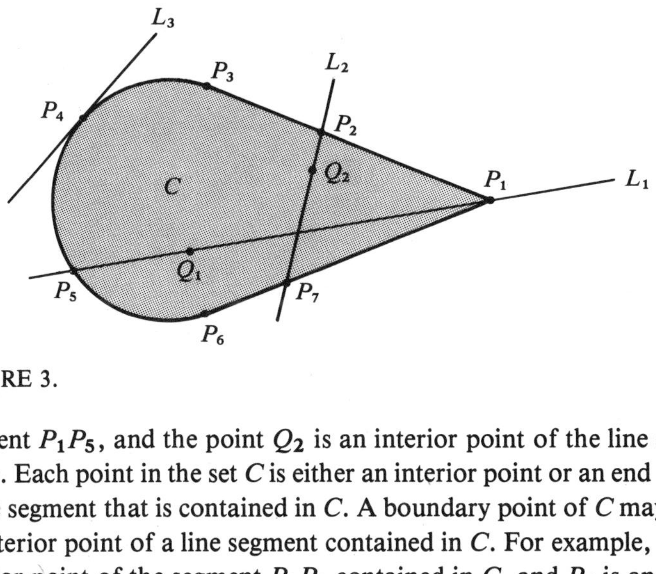

FIGURE 3.

segment P_1P_5, and the point Q_2 is an interior point of the line segment P_2P_7. Each point in the set C is either an interior point or an end point of a line segment that is contained in C. A boundary point of C may still be an interior point of a line segment contained in C. For example, P_2 is an interior point of the segment P_1P_3 contained in C, and P_7 is an interior point of the segment P_6P_1 contained in C. On the other hand, P_4 is not an interior point of any line segment contained in C. Such a point of C is called an *extreme point* of C. In Figure 3 the extreme points are P_1 and all of the points on the boundary arc $P_3P_4P_5P_6$.

We let $E(C)$ denote the set of all extreme points of C It is geometrically obvious that the set C shown in Figure 3 is the union of the points of all line segments whose end points are extreme points of C (line segments such as P_2P_7 are not needed to obtain C).

In another situation, line segments with end points in $E(C)$ may not be sufficient to generate C. Suppose that C consists of a triangle PQR plus all the points in the interior of the triangle. Then $E(C)$ contains only the three points P, Q, and R. The set of all line segments with end points in $E(C)$ will give only the sides and not the points in the interior of the triangle. To generate all of C we may take all linear combinations of the form $\alpha P + \beta Q + \gamma R$ where α, β, and γ are nonnegative numbers with $\alpha + \beta + \gamma = 1$.

As a third example consider the sector consisting of the origin plus all w for which $|\arg w| \leq \pi/4$. Here the set $E(C)$ contains only *one* point, the origin, and obviously C cannot be generated by linear combinations of points from $E(C)$. For the theorem we have in view, we must assume that C is a bounded set.

How do we generalize these simple ideas about the plane to an arbitrary set C in an arbitrary space X? First, we need the idea of a line segment. Thus our space must have operations of addition and scalar multiplication. Thus, if f and g are any points in X and α and β are any two numbers from a specified field $\mathscr{F}$, then

$$h \equiv \alpha f + \beta g$$

is also in X. A space X that has this property is called a *vector space over the field* $\mathscr{F}$. Since our main objective is the space $\mathscr{A}$ of all functions regular (analytic) in E with $\mathscr{F} = \mathscr{C}$, the set of all complex numbers, we will *not* mention $\mathscr{F}$ again. We also need the idea of a closed set and a compact set, and hence the set X must have a topology. In the set $\mathscr{A}$ the topology is that induced by uniform convergence on compact subsets of E, and we will *not* mention this item again in our applications. Finally, we need the concept of a locally convex space. A topological vector space is called *locally convex* if there is a base for the topology consisting of convex sets. With these preparations we can define convex sets and extreme points in X.

Definition 5. The line segment $L[f, g]$ joining f and g in a vector space is the set of all points h of the form $h = tf + (1 - t)g$ for which $0 \leq t \leq 1$. The points f and g are the *end points* of $L[f, g]$. The point h is an *interior point* of $L[f, g]$ if $0 < t < 1$.

Definition 6. A set C in X is *convex* if for every pair of points f, g in C, the set $L[f, g]$ is also in C.

Definition 7. A point f in a convex set C is called an *extreme point* of C if it is not an interior point of any segment $L[f, g]$ that is contained in C. We let $E(C)$ denote the set of extreme points of C.

Note that if f is an interior point of C, then it cannot be in $E(C)$. However, this does not mean that all boundary points of C are in $E(C)$; for example, in Figure 3 the points P_2 and P_7 are boundary points of C but are not extreme points of C.

If a set $\mathscr{M}$ is not convex, it can be enlarged to a convex set using

Definition 8. The *closed convex cover* (or hull) of a set $\mathscr{M}$ is the smallest closed convex set that contains $\mathscr{M}$. We denote[1] this set by $H(\mathscr{M})$, the closed hull of $\mathscr{M}$.

Such a set always exists because $\mathscr{M} \subset X$, a closed convex set, and the intersection of all closed convex sets that contain $\mathscr{M}$ is a closed convex set that contains $\mathscr{M}$.

With these preparations we are ready for

Theorem 21. PWO. Krein-Mil′man Theorem. Let C be a compact convex set in a locally convex topological vector space X. Then C is the closed convex hull of its extreme points. In symbols $C = H(E(C))$.

We omit the proof of this theorem, because it will take us too far away from the subject of this book. A proof can be found in any complete text on functional analysis, (see for example, Dunford and Schwartz [1, vol. I, p. 440]).

The solution of many problems for a given set $\mathscr{M}$ can be reduced to solving the problem for the extreme points of $\mathscr{M}$. Thus it is very helpful to find the set $E(\mathscr{M})$.

Most of the sets that arise in the theory of univalent functions are not convex. For example, CV is *not* convex (but TR is convex). However, any set $\mathscr{M}$ is always contained in its closed convex hull and frequently we have good luck: the solution of a certain extremal problem in $\mathscr{M}$ is the same as the solution in the larger set $H(\mathscr{M})$.

[1] Some authors use $c\ell(co\mathscr{M})$ for the closed hull of $\mathscr{M}$.

Our immediate objective is the proof of the Herglotz Theorem using the Krein-Mil′man Theorem. This is the main item in the next section.

8. THE HERGLOTZ REPRESENTATION FORMULA

This is

Theorem 22. Herglotz 1911 [1-B-526]. If $f(z)$ is in P, then there is a real-valued nondecreasing function $\mu(\phi)$ such that

$$\int_0^{2\pi} d\mu(\phi) = 2\pi \tag{60}$$

and for each z in E

$$f(z) = \frac{1}{2\pi}\int_0^{2\pi} \frac{1+ze^{-i\phi}}{1-ze^{-i\phi}}\,d\mu(\phi) = \frac{1}{2\pi}\int_0^{2\pi} L_0(e^{-i\phi}z)\,d\mu(\phi). \tag{61}$$

Conversely, if $f(z)$ is defined by (60) and (61) and $\mu(\phi)$ is nondecreasing, then $f(z)$ is in P.

Corollary If $f(z)$, given by

$$f(z) = 1 + \sum_{n=1}^{\infty} p_n z^n, \tag{62}$$

is in P, then for all $n \geq 1$

$$p_n = \frac{1}{\pi}\int_0^{2\pi} e^{-in\phi}\,d\mu(\phi). \tag{63}$$

We begin with a formula that is easier to prove.

Theorem 23. Let $f(z)$, given by (62), be regular in E and let $U(z)$ and $V(z)$ denote the real and imaginary parts, respectively, of $f(z)$. Set $z = re^{i\theta}$ and $\zeta = \varrho e^{i\phi}$. If $|z| < |\zeta| < 1$, then

$$f(z) = \frac{1}{2\pi}\int_0^{2\pi} U(\varrho e^{i\phi})\,\frac{\zeta+z}{\zeta-z}\,d\phi + iV(0). \tag{64}$$

This formula for $f(z)$ is often referred to as the Schwarz representation formula. It expresses $f(z)$ in terms of its real part on a slightly larger circle $|\zeta| = \varrho > |z|$.

To obtain Theorem 22 from Theorem 23, we might try to pass to the limit as $|\zeta| \to 1$. Since $f(z)$ is in P, we can set $U(e^{i\phi})d\phi = d\mu(\phi)$, because $U(e^{i\phi}) \geq 0$. Unfortunately this limit operation is not correct unless some additional hypotheses are made about $f(z)$ (see Nevanlinna [1, pp. 189–196]). We will avoid these difficulties by finding the extreme points of P and using Theorem 21.

Proof of Theorem 23. Using Cauchy's formula and integrating on the circle $|\zeta| = \varrho < 1$, we have

$$p_n = \frac{1}{2\pi i}\int_{|\zeta|=\rho} \frac{f(\zeta)}{\zeta^{n+1}}\,d\zeta = \frac{1}{2\pi\varrho^n}\int_0^{2\pi} f(\zeta)e^{-in\phi}\,d\phi, \tag{65}$$

for $n = 0, 1, \ldots.$ If we put ζ^{n-1} in the numerator, we have instead

$$0 = \frac{1}{2\pi i}\int_{|\zeta|=\rho} f(\zeta)\zeta^{n-1}\,d\zeta = \frac{\varrho^n}{2\pi}\int_0^{2\pi} f(\zeta)e^{in\phi}\,d\phi, \tag{66}$$

for $n = 1, 2, \ldots.$ We divide equation (66) by ϱ^{2n}, take the conjugate, and add the result to equation (65). This gives

$$p_n = \frac{1}{2\pi\varrho^n}\int_0^{2\pi} (f(\zeta)+\overline{f(\zeta)})e^{-in\phi}\,d\phi$$

$$p_n = \frac{1}{2\pi\varrho^n}\int_0^{2\pi} 2U(\zeta)e^{-in\phi}\,d\phi \tag{67}$$

for $n = 1, 2, \ldots.$ For each such n, we multiply (67) by z^n and add the resulting equations (including in the sum equation (65) when $n = 0$). Since $|z| < \varrho$, the convergence of the infinite series we obtain is assured. Thus

$$f(z) = p_0 + \sum_{n=1}^{\infty} p_n z^n = \frac{1}{2\pi}\int_0^{2\pi} f(\zeta)\,d\phi + \sum_{n=1}^{\infty}\frac{z^n}{2\pi\varrho^n}\int_0^{2\pi} 2U(\zeta)e^{-in\phi}\,d\phi$$

$$= iV(0) + \frac{1}{2\pi}\int_0^{2\pi} U(\zeta)\,d\phi + \frac{1}{2\pi}\int_0^{2\pi} U(\zeta)\left(\sum_{n=1}^{\infty} 2\,\frac{z^n}{\zeta^n}\right)\,d\phi$$

$$f(z) = \frac{1}{2\pi}\int_0^{2\pi} U(\zeta)\left[1 + 2\sum_{n=1}^{\infty}\frac{z^n}{\zeta^n}\right]\,d\phi + iV(0). \tag{68}$$

Since the series in (68) yields $(\zeta + z)/(\zeta - z)$, this equation gives (64). ■

If $p_0 = 1$, then $iV(0) = 0$.

We next find the extreme points in P.

Theorem 24. The function $E(z)$ is an extreme point in P if and only if

(69) $$E(z) = \frac{1+\eta z}{1-\eta z},$$

where η is a complex number with $|\eta| = 1$.

Proof. We give here the proof due to Clunie and Holland 1981 [1]. We first recall that if $f(z) = 1 + p_1 z + \cdots$ is in P, then $|p_1| \leq 2$. Further, in Problem 5 we will see that $|p_1| = 2$ if and only if $f(z)$ has the form (69).

If $E(z)$ has the form (69), then $E(z)$ is an extreme point. Suppose to the contrary that $E(z)$ is not in $E(P)$. Then we have two functions, $f(z)$ and $g(z)$, in P such that

$$E(z) = tf(z) + (1-t)g(z) = t(1 + a_1 z + \cdots) + (1-t)(1 + b_1 z + \cdots)$$

so that

(70) $$2|\eta| = 2 \leq t|a_1| + (1-t)|b_1|, \qquad 0 < t < 1.$$

Since $|a_1| \leq 2$ and $|b_1| \leq 2$, this is possible if and only if $|a_1| = |b_1| = 2$ and $\operatorname{Arg} a_1 = \operatorname{Arg} b_1$. But then both $f(z)$ and $g(z)$ have the form (69) and, in fact, $f(z) = g(z)$. Hence $E(z)$ is an extreme point of P.

It is harder to prove that we now have all the extreme points in P. If $f(z)$ does not have the form (69), and $f(z) = 1 + p_1 z + p_2 z^2 + \cdots$, then $|p_1| < 2$. We will show that in this case $f(z)$ is an interior point of the line joining two functions $g(z)$ and $h(z)$ in P. This will prove that $f(z)$ is not an extreme point.

First, we select c so that $\bar{c}p_1 = |p_1|$. Then $|c| = 1$. If $p_1 = 0$, then we can set $c = 1$. We next construct the auxiliary function

(71) $$F(z, c) \equiv \frac{f(z)(1-cz)(z-\bar{c})}{z} + \frac{\bar{c}}{z} - cz$$

$$= -p_1\bar{c} + 1 + |c|^2 + [-p_2\bar{c} + p_1 + |c^2|p_1 - 2c]z + \cdots$$

(72) $$= -p_1\bar{c} + 2 + [-p_2\bar{c} + 2p_1 - 2c]z + \cdots.$$

We observe that the constant term $2-p_1\bar{c} = 2-|p_1| > 0$, because $|p_1| < 2$. Now suppose that $f(z)$ is regular on ∂E. Then for $z = e^{i\theta}$

$$\begin{aligned}\operatorname{Re} F(e^{i\theta}, c) &= \operatorname{Re}\,[f(e^{i\theta})(2 - ce^{i\theta} - \bar{c}e^{-i\theta}) + \bar{c}e^{-i\theta} - ce^{i\theta}]\\ &= \operatorname{Re}\,[2 - ce^{i\theta} - \bar{c}e^{-i\theta}]\operatorname{Re} f(e^{i\theta}) \geq 0.\end{aligned}$$

Since $F(z, c)$ is regular in $\bar{E}$, this inequality holds throughout E. If $f(z)$ has singularities on ∂E, we apply the work just completed to $f(rz)$ and let $r \to 1^-$. Hence if $f(z)$ is in P, then $\operatorname{Re} F(z, c) > 0$ in E. Further, each step from equation (71) on is valid if we replace c by $-c$. We define $g(z)$ and $h(z)$ by

$$g(z) = \frac{F(z, c)}{2-|p_1|} \quad \text{and} \quad h(z) = \frac{F(z, -c)}{2+|p_1|}. \tag{73}$$

Clearly $g(0) = h(0) = 1$ (see equation (72)), so that both $g(z)$ and $h(z)$ are in P. A brief computation shows that

$$f(z) = \frac{2-|p_1|}{4}g(z) + \frac{2+|p_1|}{4}h(z) \tag{74}$$

where $0 < (2 \pm |p_1|)/4 < 1$. Thus, from (74) it appears as though $f(z)$ is not an extreme point in P. There is one minor detail to check. It is possible that $g(z)$ and $h(z)$ are the same function, and if so the argument fails. Assume that $g(z) = h(z)$. Then a moderately long computation will give

$$f(z) = \frac{1-c^2z^2}{1-|p_1|cz + cz^2}, \tag{75}$$

and if $|p_1| = 2\cos\beta$, then $|\cos\beta| < 1$ and

$$f(\bar{c}z) = \frac{1}{2}\left(\frac{1+e^{i\beta}z}{1-e^{i\beta}z} + \frac{1+e^{-i\beta}z}{1-e^{-i\beta}z}\right). \tag{76}$$

Again, $f(z)$ is not an extreme point in P. ∎

Theorem 24 is very important, because it will help us to find extreme points in other sets.

Proof of Theorem 22. We apply the Krein-Mil'man Theorem, where now the space X is $\mathscr{A}$ and the convex set C is P. To say that $C =$

$H(E(C))$, or in this case $P = H(E(P))$, is to say that any function in P can be obtained as a weighted average[1] of the extreme points in P. But this is exactly what the Herglotz formula states (now that we know all the extreme points in P). ■

Komatu 1958 [1-B-718 and 720] gave an extension of the Herglotz representation formula to functions that have positive real part in a ring domain. See also Nishimiya 1959 [3-B-640].

Theorem 25. Let PR be the subset of P of functions that are real in $(-1, 1)$. Then there is a real-valued nondecreasing $\mu(\phi)$ with the normalization (60) such that for z in E

$$f(z) = \frac{1}{2\pi} \int_0^{2\pi} \frac{1 - z^2}{1 - 2z \cos \phi + z^2} \, d\mu(\phi). \tag{77}$$

Proof. Since $f(z)$ is real for $-1 < z < 1$, the coefficients p_n are all real. Thus $f(z) = \overline{f(\bar{z})}$, or

$$f(z) = \frac{f(z) + \overline{f(\bar{z})}}{2} = \frac{1}{4\pi} \int_0^{2\pi} \left(\frac{1 + e^{-i\phi} z}{1 - e^{-i\phi} z} + \frac{1 + e^{i\phi} z}{1 - e^{i\phi} z} \right) d\mu(\phi).$$

This gives equation (77). ■

PROBLEMS

1. Prove that $L_0(z) \equiv (1 + z)/(1 - z)$ has the power series (3).
2. Prove that $L_0(E)$ is the right half-plane, and that $L_0(z)$ is univalent in E.
3. Prove Theorem 2.
4. ★ Referring to Theorem 3, prove that if $p_N = 2$, then $F(z)$ has the form given in equations (15) and (16). Hint: Use the Stieltjes integral representation given in equation (63).
5. If $f(z) = 1 + p_1 z + p_2 z^2 + \cdots$ is in P and $|p_1| = 2$, prove that $f(z) = (1 + \eta z)/(1 - \eta z)$ with $|\eta| = 1$. Hint: Apply the second part of Schwarz's Lemma to $g(z) = (f(z) - 1)/(f(z) + 1)$. As a generalization prove that if $f(z) \prec F(z)$, then $|a_1| \le A_1$ with

[1] According to Schober, this passage from $H(E(P))$ to an integral is not so simple and requires Choquet's Theorem. See Schober 1975 [2-B-1280, pp. 4 and 173–175].

equality if and only if $f(z) = F(\eta z)$ with $|\eta| = 1$. Here $a_1 = f'(0)$ and $A_1 = F'(0)$.

6. Prove Theorems 5 and 6.
7. Let P' be the set of functions $f(z) = z + \Sigma_2^\infty a_n z^n$ for which $f'(z)$ is in P. Prove that $|a_n| \leq 2/n$. Is this inequality sharp? What are the extremal functions? Functions in P' are often called *functions of bounded turning*, because $\beta \equiv \arg f'(z_0)$ is the angle of turn of the image of a line segment at z_0 under $f(z)$. For functions in P' we have $|\beta| < \pi/2$. See Zmorovič 1950 [1-B-1516] and Gal'perin 1958 [1-B-350].
8. If $f(z)$ is in P', describe a domain in which $f'(E_r)$ must lie.
9. Use the result of Problem 8 and integration to obtain sharp bounds for $|f(z)|$ when $f(z)$ is in P'.
10. Suppose that $f(z) << F(z)$ in E. Prove that for each r in $[0, 1)$ the area of $f(E_r)$ is less than or equal to the area of $F(E_r)$.
11. Let $f(z) = a + bi + \Sigma_1^\infty a_n z^n$ be regular in E and suppose further that $\text{Re}\, f(z) > \alpha$ in E, where $\alpha < a$. Find sharp bounds for $|a_n|$ and at least one extremal function.
12. If $f(z)$ satisfies the conditions of Problem 11, find the smallest disk that will always contain $f(E_r)$.
13. Let $f(z) = z + \Sigma_2^\infty a_n z^n$ be regular in E and suppose further that $|\arg f'(z)| < \beta\pi/2$ in E where $\beta < 1$. Prove that

$$|f'(z)| \leq [(1+r)/(1-r)]^\beta.$$

14. Under the conditions of Problems 13, prove that $|f(z)|$ is bounded in E, and find an acceptable bound. What is the situation if $\beta \geq 1$? Rogosinski 1923 [1-B-1177].
15. Suppose that $f(z) = \Sigma_0^\infty a_n z^n$ is regular in E and $|\text{Re}\, f(z)| < \delta/2$ in E. Prove that

$$\sum_{n=1}^{\infty} |a_n|^2 \leq \frac{\delta^2}{2},$$

and that this inequality is sharp. Tammi 1957 [1-B-1382].
16. Prove that if $f(E)$ lies in a strip of width δ, then the inequality of Problem 15 is still true. Thus, the inequality does not depend on the location of the strip. Tammi [1-B-1382]. For some generalizations see MacGregor 1969 [2-B-907] and Robertson 1978 [3-B-756].

17. Let $f(z) = \Sigma_{-\infty}^{\infty} a_n z^n$ be regular in the ring domain $E(q, 1)$: $q < |z| < 1$. If $\text{Re } f(z) > 0$ in $E(q, 1)$, prove that

$$|a_n + \bar{a}_{-n}| \leq 2 \text{ Re } a_0.$$

Observe that the bound does not depend on q. Hint:

$$a_n r^n = \frac{1}{2\pi} \int_0^{2\pi} f(re^{i\theta}) e^{-in\theta} \, d\theta, \qquad q < r < 1.$$

Robertson 1939 [1-B-1151].

★ **18.** Let $f(z)$ satisfy the conditions of Problem 17. Prove that for $n = 1, 2, 3, \ldots$

$$|a_n| \leq \frac{2}{1 - q^n} \text{ Re } a_0, \qquad |a_{-n}| \leq \frac{2q^n}{1 - q^n} \text{ Re } a_0.$$

An example function that shows these bounds are sharp requires the use of Jacobian elliptic functions. Nehari 1950 [1-B-974]. For further work see Nishimiya 1959 [3-B-640].

19. Prove that if $f(z)$ is in P, then

$$f(z) = \frac{1}{z} \int_0^z f(\zeta) \, d\zeta$$

is also in P.

20. Use equation (4) with suitable functions to prove that

$$g(z) \equiv \frac{1 - z^2}{1 - 2z \cos \alpha + z^2}$$

is in P for every real α. Rogosinski 1932 [1-B-1180].

21. Use equation (11) with suitable f_1, f_2, t_1 and t_2, to prove that

$$g(z) = \frac{1 - z}{\sqrt{1 - 2z \cos \alpha + z^2}}$$

is in P for every real α. Rogosinski [1-B-1180].

22. Recall that if $P_n(x)$ is the nth Legendre polynomial, then

$$\frac{1}{\sqrt{1 - 2z \cos \alpha + z^2}} = 1 + \sum_{n=1}^{\infty} P_n(\cos \alpha) z^n.$$

Prove that if $g(z)$ is the function defined in Problem 21, then

$$g(z) = 1 + \sum_{n=1}^{\infty} [P_n(\cos\alpha) - P_{n-1}(\cos\alpha)]z^n.$$

23. Find $g(E)$ for the function defined in Problem 21. See Fejér 1925 [1-B-292] for Problems 23 through 27.

24. Let $\sigma_n(z) = \dfrac{n+1}{2} + nz + (n-1)z^2 + \cdots + 2z^{n-1} + z^n$. Use equation (1.11) to prove that $\operatorname{Re}\sigma_n(z) > 0$ in E.

25. Let $f(z) = \dfrac{c_0}{2} + \sum_{k=1}^{\infty} c_k z^k$ and let $S_n(z)$ be the nth partial sum:

$$S_n(z) = \frac{c_0}{2} + \sum_{k=1}^{n} c_k z^k \qquad n = 0, 1, 2, \ldots.$$

With the notation of Problem 24 prove that

$$S_n(z) = c_n\sigma_n(z) + (c_{n-1} - 2c_n)\sigma_{n-1}(z) + \sum_{k=0}^{n-2} (c_k - 2c_{k+1} + c_{k+2})\sigma_k(z).$$

26. A sequence $a_0, a_1, \ldots, a_n, \ldots$ of nonnegative numbers is called a *convex null sequence* if $a_n \to 0$ as $n \to \infty$ and

$$a_0 - a_1 \geq a_1 - a_2 \geq \cdots \geq a_n - a_{n+1} \geq \cdots \geq 0.$$

Prove that if $a_0, a_1, \ldots, a_n, \ldots$ and $b_0, b_1, \ldots, b_n, \ldots$ are two convex null sequences, then $a_0b_0, a_1b_1, \ldots, a_nb_n, \ldots$ is a convex null sequence. Prove that if $0 < r < 1$, then $a_0, a_1r, a_2r^2, \ldots, a_nr^n, \ldots$ is also a convex null sequence.

27. Suppose that $c_0, c_1, \ldots, c_n, \ldots$ is a convex null sequence and

$$g(z) = \frac{c_0}{2} + \sum_{k=1}^{\infty} c_k z^k.$$

Use the results of Problems 24, 25, and 26 to prove that $\operatorname{Re} g(z) > 0$ for z in E.

28. Let $f(z)$ be in S and suppose that we know that $|a_n| \leq n$ for all n. Use the technique of dominant power series to prove that if $n \geq 2$ and z is in E, then

$$|f^{(n)}(z)| \leq n! \, \frac{n+r}{(1-r)^{n+2}}\,.$$

Note that if $n = 2$ or 3, this inequality can be proved by other means. See Problem 6.14.

For Problems 29 through 34 see Robertson 1981 [3-B-758].

29. Prove that if $p(z) = 1 + \Sigma_1^\infty p_n z^n$ is in P, then

$$q(z) \equiv \frac{1-z^2}{z} - \frac{(1-z)^2}{z} p(z) \tag{78}$$

is regular in E and Re $q(z) \geq 0$ in E.

30. Prove that for $p(z)$ and $q(z)$ defined in Problem 29 we have

$$q(z) = 2 - p_1 + (2p_1 - p_2 - 2)z + \sum_{n=2}^{\infty} (2p_n - p_{n-1} - p_{n+1})z^n.$$

31. What coefficient theorem can you derive from the work in Problems 29 and 30?

32. Under the conditions of Problem 29 prove that for $n \geq 1$ we have

$$|p_{n+1} - p_n| \leq (2n+1)|2 - p_1|$$

and

$$||p_{n+1}| - |p_n|| \leq (2n+1)(2 - |p_1|).$$

Hint: Put (78) in the form $(1-z)p(z) = 1 + z - zq(z)/(1-z)$.

33. In Problem 29 find $q(z)$ if $p(z) = (1+z)/(1-z)$. Is there an invariant function under the transformation (78)?

34. Prove that $\dfrac{1-z^2}{1 - 2z\cos\phi + z^2} = 1 + \displaystyle\sum_{n=1}^{\infty} 2(\cos n\phi)z^n.$

Hint: Use partial fractions. See the proof of Theorem 25.

★ **35.** Show that for each n the inequality proved in Problem 32 is *best possible*. Note that this is not quite the same as *sharp*. In a sharp inequality there is some admissible function in which equality holds. If an inequality is best possible, equality may not hold but any decrease in the larger side will make the assertion false. Hint: Use the example function from Problem 34 and let $\phi \to 0$, so that $p_1 \to 2$. If we consider the subset of P with p_1 fixed, then the sharp upper bound for $|p_{n+1} - p_n|$ is not known.

36. Prove that if $p(z)$ is in P, then

$$\left|\frac{zp'(z)}{p(z)}\right| \leq \frac{2r}{1-r^2} \tag{79}$$

Is this inequality sharp? Hint: Consider $p(L(z))$. Goluzin 1935 [1-B-383, p. 186]. For a generalization see Bernardi 1974 [2-B-147].

37. For $0 \le \alpha < 1$, let $P(\alpha)$ denote the subset of P of functions for which $\operatorname{Re} p(z) > \alpha$ in E. Prove that if $p(z)$ is in $P(\alpha)$, then there is a $b(z)$ in B_0 such that

$$p(z) = \frac{1 + (1 - 2\alpha)b(z)}{1 - b(z)}. \tag{80}$$

Find a Stieltjes integral representation for functions in $P(\alpha)$. See Problem 11 for the sharp upper bound for $|p_n|$.

38. Let $\widetilde{P}(M)$ be the subset of P of functions for which $|p(z) - M| < M$ for z in E, where $M > 1/2$. Find a formula for $p(z)$ similar to equation (80). Goel 1967 [2-B-440], [2-B-443], and [2-B-446].

39. Set $m = 1 - 1/M$ for $M > 1/2$. Prove that

$$P(z) = \frac{1+z}{1-mz} = 1 + \sum_{n=1}^{\infty} m^{n-1}(1+m)z^n$$

is in $\widetilde{P}(M)$. This suggests the conjecture $|p_n| \le m^{n-1}(1+m)$ for $p(z)$ in $\widetilde{P}(M)$. Does this conjecture seem reasonable?

40. Let $B(z)$ satisfy the conditions of Schwarz's Lemma and suppose further that $B(1) = 1$. Prove that

$$|1 - B(z)|/(1 - |z|) \ge 1$$

for all z in E.

41. Let $B(z)$ satisfy the conditions of Problem 40, and suppose further that $B(z)$ is regular at $z = 1$. Prove that $|B'(1)| \ge 1$.

42. Referring to the sets $P(\alpha)$ and $\widetilde{P}(M)$ defined in Problems 37 and 38, respectively, prove that $p(z)$ is in $P(\alpha)$ if and only if $1/p(z)$ is in $\widetilde{P}(M)$ where $M = 1/2\alpha$. Here $0 < \alpha < 1$. Thus any theorem proved about one of the sets usually gives a companion theorem for the other set.

43. Let $A = 1 - 2\alpha$, so that $0 \le \alpha \le 1$ if and only if $-1 \le A \le 1$. Prove that if $f(z)$ is in $P(\alpha)$, then

$$\frac{1 - Ar}{1 + r} \le \operatorname{Re} p(z) \le |p(z)| \le \frac{1 + Ar}{1 - r}. \tag{81}$$

Watch out for the case $-1 \le A < 0$.

44. Set $m = 1 - 1/M$. Prove that if $M > 1/2$ and $p(z)$ is in $\widetilde{P}(M)$ then

$$\frac{1 - r}{1 + mr} \le \operatorname{Re} p(z) \le |p(z)| \le \frac{1 + r}{1 - mr}. \tag{82}$$

45. For fixed $n \geq 1$, assume that $p(z) = 1 + \Sigma_{k=n}^{\infty} a_k z^k = 1 + a_n z^n + \cdots$ is in $\widetilde{P}(M)$. Prove that if $m = 1 - 1/M$ is in $(-1, 1)$, then

$$\frac{1 - r^n}{1 + mr^n} \leq \operatorname{Re} p(z) \leq |p(z)| \leq \frac{1 + r^n}{1 - mr^n}. \tag{83}$$

46. Let $P^{(k)}$ denote the set of functions of the form

$$p(z) = 1 + p_k z^k + p_{2k} z^{2k} + \cdots = 1 + \sum_{n=1}^{\infty} p_{kn} z^{kn} \tag{84}$$

for which $\operatorname{Re} p(z) > 0$ in E. For brevity we say that $p(z)$ has *k-fold symmetry*. Find a Stieltjes integral representation for functions in $P^{(k)}$.

47. For fixed r find the domain in which $p(E_r)$ must lie if $p(z)$ is in $P^{(k)}$.

CHAPTER 8

CONVEX AND STARLIKE FUNCTIONS

1. DEFINITIONS

If we know that a univalent function maps E onto a domain with some nice property, then we have the means for a more penetrating study of the function. A convex domain is an outstanding example of a domain with nice properties. Another example is a domain that is starlike with respect to some point.

Although our interest centers on open sets, we frame the definitions so that they include regions (domains plus none, some, or all of the boundary points).

Definition 1. A set D in the plane is called *convex* if for every pair of points w_1 and w_2 in the interior of D, the line segment joining w_1 and w_2 is also in the interior of D. If a function $f(z)$ maps E onto a convex domain, then $f(z)$ is called a *convex function*.

Definition 2. A set D in the plane is said to be *starlike with respect to* w_0 an interior point of D if each ray with initial point w_0 intersects the interior of D in a set that is either a line segment or a ray. If a function $f(z)$ maps E onto a domain that is starlike with respect to w_0, then we say that $f(z)$ *is starlike with respect to* w_0. In the special case that $w_0 = 0$, we say that $f(z)$ is a *starlike function.*

Examples of convex and starlike domains are shown in Figures 1 and 2, respectively. The domain shown in Figure 2 is starlike with respect to w_0 but it is *not* starlike with respect to the origin. The special function $L_0(z) = (1+z)/(1-z)$ is a convex function because it maps E onto a half-plane. The Koebe function is a starlike function. In fact the domain $k(E)$ is starlike with respect to each $w_0 > -1/4$.

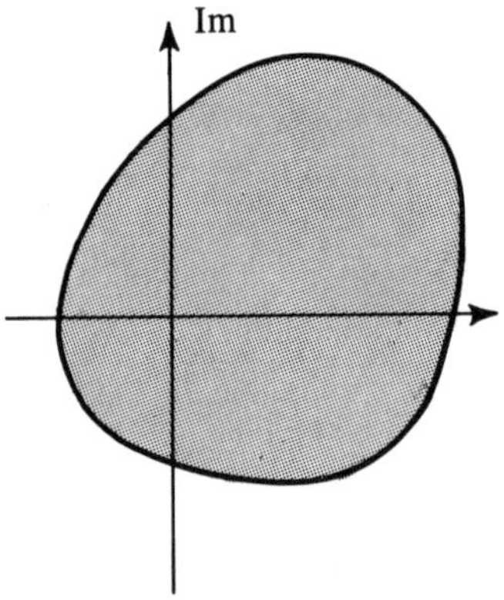

FIGURE 1

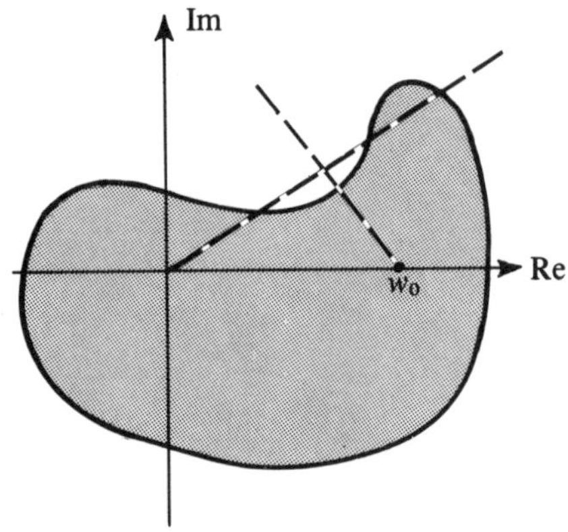

FIGURE 2

Any circular disk or any half-plane is a convex set. The intersection of any number of convex sets is a convex set (of course, the intersection may be empty or consist of only one point). A convex set D is starlike with respect to each interior point of D. Conversely, if a set D is starlike with respect to each interior point of D, then D is a convex set.

2. CONVEXITY AND STARLIKENESS ON A CURVE

We begin our study of convex and starlike functions by considering the image of a curve Γ_z under a function $f(z)$ that is analytic on Γ_z. Eventually Γ_z will be a circle, a line segment, or some other elementary arc. Hence we assume that Γ_z is a smooth curve with a parametrization

$$z(t) = x(t) + iy(t), \qquad a \le t \le b \tag{1}$$

where $x(t)$ and $y(t)$ are real functions, and

$$z'(t) = x'(t) + iy'(t) \neq 0, \tag{2}$$

for t in $[a, b]$. The arc Γ_z is a directed arc, the direction being that determined as t increases.

Let Γ_w be the image of Γ_z under a function $f(z)$ that is analytic on Γ_z, and assume that w_0 is not on Γ_w (see Figure 3). The arc Γ_w is said to be starlike with respect to w_0 if $\arg(w - w_0)$ is a nondecreasing function[1] of t, i.e., if

$$\frac{d}{dt} \arg(w - w_0) \ge 0, \qquad t \text{ in } [a, b]. \tag{3}$$

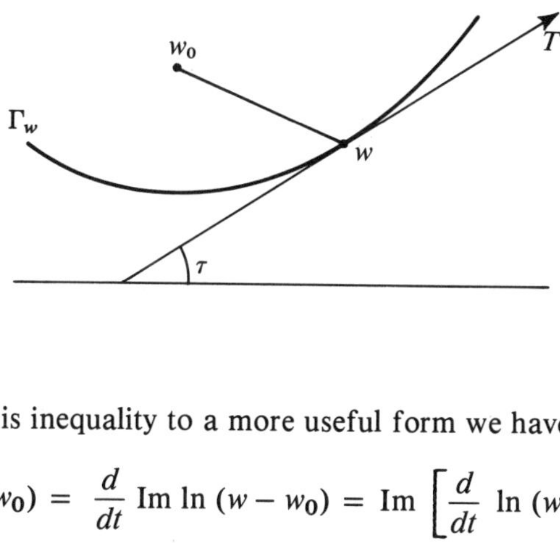

FIGURE 3

To convert this inequality to a more useful form we have

$$\frac{d}{dt} \arg(w - w_0) = \frac{d}{dt} \operatorname{Im} \ln(w - w_0) = \operatorname{Im}\left[\frac{d}{dt} \ln(w - w_0)\right]$$
$$= \operatorname{Im}\left[\frac{d}{dz} \ln(w - w_0)\frac{dz}{dt}\right] = \operatorname{Im}\left[\frac{f'(z)}{f(z) - w_0}\,\frac{dz}{dt}\right]$$

[1] Of course, $\arg(w - w_0)$ is a multiple-valued function, but under the given conditions we can select a single-valued branch that is differentiable. Henceforth, we will always make such a selection.

Lemma 1. The image of Γ_z: $z = z(t)$ under $f(z)$ is starlike with respect to w_0 if and only if

$$\text{(4)} \qquad \operatorname{Im}\left[\frac{f'(z)}{f(z)-w_0}\, z'(t)\right] \geq 0, \qquad t \text{ in } [a, b].$$

The arc Γ_w is said to be convex if the argument of the tangent to Γ_w is a nondecreasing function of t (see Figure 3). The direction of the tangent to Γ_z is $\arg z'(t)$ and the mapping rotates this tangent vector through an angle $\arg f'(z)$. Thus the arc Γ_w is a convex arc if and only if

$$\text{(5)} \qquad \frac{d\tau}{dt} = \frac{d}{dt}\left[\arg\left(z'(t) f'(z)\right)\right] \geq 0, \qquad t \text{ in } [a, b].$$

The same technique used on equation (3) gives

$$\begin{aligned}
\frac{d}{dt}\left[\arg\left(z'(t) f'(z)\right)\right] &= \frac{d}{dt} \operatorname{Im}\left[\ln z'(t) + \ln f'(z)\right] \\
&= \operatorname{Im}\left[\frac{z''(t)}{z'(t)} + \frac{d}{dz}\left(\ln f'(z)\right)\frac{dz}{dt}\right] \\
&= \operatorname{Im}\left[\frac{z''(t)}{z'(t)} + \frac{f''(z)}{f'(z)}\, z'(t)\right].
\end{aligned}$$

Lemma 2. Suppose that $f'(z) \neq 0$ on Γ_z: $z = z(t)$. Then the image of Γ_z under $f(z)$ is a convex arc if and only if

$$\text{(6)} \qquad \operatorname{Im}\left[\frac{z''(t)}{z'(t)} + \frac{f''(z)}{f'(z)}\, z'(t)\right] \geq 0, \qquad t \text{ in } [a, b].$$

We now specialize these formulas by selecting Γ_z to be the circle C_R: $|z| = R$. Thus with the usual orientation, $z = Re^{it}$, $0 \leq t \leq 2\pi$. In this case $z'(t) = iRe^{it} = iz$ and $z''(t) = -Re^{it} = -z$. The inequality (4) becomes

$$\text{(7)} \qquad \operatorname{Im}\left[i\,\frac{zf'(z)}{f(z)-w_0}\right] = \operatorname{Re}\left[\frac{zf'(z)}{f(z)-w_0}\right] \geq 0,$$

and the inequality (6) becomes

$$\text{(8)} \qquad \operatorname{Im}\left[i + i\,\frac{zf''(z)}{f'(z)}\right] = \operatorname{Re}\left[1 + \frac{zf''(z)}{f'(z)}\right] \geq 0,$$

for z on C_R.

3. FUNCTIONS UNIVALENT IN THE UNIT DISK

Now let $f(z)$ be regular and univalent in E. If $f(z)$ maps C_R onto a simple closed convex curve, then this curve bounds a convex region. Conversely, if $f(z)$ maps the disk E_R, $R < 1$, onto a convex region, then the boundary of the region is a simple closed convex curve. A similar statement holds for starlike curves. Using equations (7) and (8) we have the following theorem (where for simplicity we set $w_0 = 0$).

Theorem 1. Let $f(z)$ be regular and univalent in the closed disk $\bar{E}_R$: $|z| \leq R$. Then $f(z)$ maps $\bar{E}_R$ onto a convex domain if and only if

$$\text{(9)} \qquad \operatorname{Re}\left[1 + \frac{zf''(z)}{f'(z)}\right] \geq 0, \qquad \text{for } z \text{ on } C_R\colon |z| = R.$$

Suppose further that $f(0) = 0$. Then $f(z)$ maps E_R onto a region that is starlike with respect to $w = 0$ if and only if

$$\text{(10)} \qquad \operatorname{Re}\left[\frac{zf'(z)}{f(z)}\right] \geq 0, \qquad \text{for } z \text{ on } C_R\colon |z| = R.$$

The condition (9) for convexity was first stated by Study 1913 [1-B-1341, p. 109]. The condition (10) for starlikeness is due to R. Nevanlinna 1921 [1-B-990].

In Theorem 1 we must assume that $f(z)$ is univalent (or replace this with some other condition) or we fall into error. Indeed, suppose that $f(z) = z^2$. Then the inequality (9) becomes $2 \geq 0$, and (10) also becomes $2 \geq 0$. Thus the inequalities (9) and (10) are satisfied, but the two-sheeted image of E under $f(z) = z^2$ is not really a convex or starlike domain in accordance with definitions 1 and 2. The concepts of convexity and starlikeness can be extended to multi-sheeted regions, and indeed these extensions have been thoroughly explored, but for the present we consider only plane regions.

We observe that if $f(z)$ is univalent in E, then $f'(z) \neq 0$ in E and hence in (9) the expression on the left is a harmonic function in E_R and takes its minimum on the boundary C_R. Thus, if $f(z)$ maps C_R onto a closed convex curve, then for each $r < R$, $f(z)$ maps C_r onto a convex curve, and hence maps $\bar{E}_r$ onto a convex domain. The same type of reasoning can be applied to the inequality (10) because if $f(z)$ is in S, then the singularity at $z = 0$ is a removable singularity. We have proved

Theorem 2. Let $f(z)$ be in S and suppose that $R < 1$. If $f(z)$ maps E_R onto a convex domain, then $f(z)$ maps E_r onto a convex domain for each positive $r < R$.

If $f(z)$ maps E_R onto a domain starlike with respect to the origin, then $f(z)$ maps E_r onto a domain starlike with respect to the origin for each positive $r < R$.

4. FORD'S THEOREM

It is easy to conjecture that Theorem 2 can be extended to the unit disk. We observe that in our approach through the inequalities (4), (6), (9), and (10) we assume that $f(z)$ is analytic on C_R ($R < 1$). However, it is possible to have a function $f(z)$ that is univalent in E but is not analytic at any point of the boundary of E (see Hurwitz [3-B-379, p. 469 and p. 478] and the references in his footnote 16).

For convex regions the extension of Theorem 2 to the unit disk was first obtained by Study [1-B-1341] by a long and detailed analysis. For starlike regions the extension was achieved independently by Takahashi 1930 [1-B-1366] and Seidel 1931 [1-B-1278].

T. Rado 1929 [1-B-1088] showed that Study's Theorem is an easy consequence of Schwarz's Lemma. Finally, L. R. Ford 1935 [1-B-320] used Rado's method to prove a theorem that contains the results of Study and Seidel, and other results as well. A few preparations are needed.

As usual we suppose that $f(z)$ is in S and $f(E) = D$. Further, suppose that $T(w_1, w_2, \ldots, w_k)$ is a function that is analytic in each variable w_j when w_j is in D. As examples, if t is fixed with $0 < t < 1$, then

(11) $$T_1(w_1, w_2) = tw_1 + (1 - t)w_2$$

is one such function, and

(12) $$T_2(w_1) = tw_1$$

is another.

For a fixed function T, we say that the domain D has the property T if $w_1, w_2, \ldots, w_k$ in D implies that

(13) $$w_0 = T(w_1, w_2, \ldots, w_k)$$

is also in D. Clearly, the function T_1 is the one associated with a convex domain. This, if D is a convex domain, it has the property T_1 for each t

in [0, 1]. Similarly, if D is starlike with respect to the origin, then D has the property T_2 for each t in [0, 1].

It should be noted that if t is fixed in (12) then the domain need not be starlike. For example, the domain shown in Figure 4 has the property T where $T(w_1) = w_1/2$ but D is not starlike.

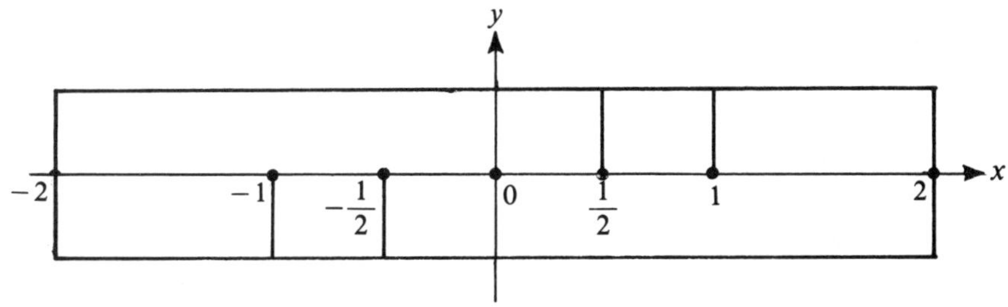

FIGURE 4.

Theorem 3. Ford. Let $f(z)$ be in S and let $D = f(E)$ have the property T, where $T(w_1, w_2, \ldots, w_k)$ is analytic in each variable w_j when w_j is in D. Further suppose that $T(0, 0, \ldots, 0) = 0$. Then for each $r < 1$ the subdomain $D_r \equiv f(E_r)$ also has the property T.

Proof. For fixed $r < 1$ let $w_1, w_2, \ldots, w_k$ be arbitrary points in $D_r = f(E_r)$. We must prove that $w_0 \equiv T(w_1, w_2, \ldots, w_k)$ is also in D_r.

Let $z = g(w)$ be the inverse function of $f(z)$, and let $z_j = g(w_j)$ for $j = 1, 2, \ldots, k$. By hypothesis $|z_j| < r$. Let $z^\star$ be some point in E_r such that

$$|z_j| < |z^\star| < r, \qquad j = 1, 2, \ldots, k, \tag{14}$$

and consider the function $b(z)$ defined in E by

$$b(z) \equiv g\left[T\left(f\left(\frac{z_1}{z^\star} z\right), f\left(\frac{z_2}{z^\star} z\right), \ldots, f\left(\frac{z_k}{z^\star} z\right)\right)\right]. \tag{15}$$

Since $T(w_1, w_2, \ldots, w_k)$ is analytic in D, and D has the property T, the function $b(z)$ is analytic in E and further $|b(z)| < 1$ for z in E (since $g(w)$ carries D onto E). Further, $b(0) = g[T(0, 0, \ldots, 0)] = 0$. Thus $b(z)$ satisfies the conditons of Schwarz's Lemma. If we set $z = z^\star$, and $z_0 = b(z^\star)$, then

$$|z_0| = |h(z^\star)| = |g[T(f(z_1), f(z_2), \ldots, f(z_k))]| \leq |z^\star| < r.$$

But $z_0 = h(z^\star)$ and $g(T(w_1, w_2, \ldots, w_k)) = g(w_0)$. Thus $w_0 = f(z_0)$ and hence w_0 is in $f(E_r)$. ■

Theorem 4. Let $f(z)$ be in S. If $f(E)$ is a convex domain, then for each positive $r < 1$, $f(E_r)$ is also a convex domain. If $f(E)$ is starlike with respect to the origin, then for each positive $r < 1$, $f(E_r)$ is also starlike with respect to the origin.

Proof. For convexity (Study's Theorem) we apply Theorem 3 with $T(w_1, w_2) = tw_1 + (1-t)w_2$ for each fixed t in $[0, 1]$. For starlikeness (the Takahashi-Seidel Theorem) we apply Theorem 3 with $T(w_1) = tw_1$ for each fixed t in $[0, 1]$. ■

5. ALEXANDER'S THEOREM

There is an elementary and beautiful relationship between convex and starlike functions that was first noticed by J. W. Alexander. Indeed, for any $f(z)$ we set

(16) $$F(z) = zf'(z).$$

Then

$$\frac{zF'(z)}{F(z)} = z\,\frac{f'(z) + zf''(z)}{zf'(z)},$$

and hence whenever $F(z) \neq 0$

(17) $$\frac{zF'(z)}{F(z)} = 1 + \frac{zf''(z)}{f'(z)}.$$

If $f(z)$ has a zero of kth order at the origin, then both sides of (17) are regular at the origin.

Now compare (17) with the inequalities (9) and (10) for convex and starlike functions. Since

(18) $$\operatorname{Re}\left[\frac{zF'(z)}{F(z)}\right] = \operatorname{Re}\left[1 + \frac{zf''(z)}{f'(z)}\right]$$

we have immediately

Theorem 5. Alexander 1915 [1-B-52]. Suppose that $f'(z) \neq 0$ in E_R. Then $f(z)$ is convex in E_R if and only if $F(z) \equiv zf'(z)$ is starlike in E_R.

Given $f(z)$, equation (16) defines $F(z)$. But we can start with an $F(z)$ that is starlike and by an integration,

$$f(z) = \int_0^z \frac{F(\zeta)}{\zeta}\, d\zeta, \tag{19}$$

find an $f(z)$ that is convex in the same disk.

As an interesting example consider

$$f(z) = \frac{z}{1-z} = z + \sum_{n=2}^{\infty} z^n, \tag{20}$$

which maps E onto the half-plane $\mathrm{Re}\, w > -1/2$. Since this function is convex in E, the function

$$F(z) \equiv zf'(z) = z\,\frac{(1-z)-z(-1)}{(1-z)^2} = \frac{z}{(1-z)^2} \tag{21}$$

is starlike in E. We recognize that the right side of (21) is the Koebe function.

Definition 3. We let CV denote the set of all normalized functions that are univalent in E and map E onto a convex domain. We let ST denote the set of all normalized functions that are univalent in E and map E onto a domain that is starlike with respect to the origin.

With this new notation we can rephrase Theorem 5 as follows:

If $f(z)$ is in CV, then $zf'(z)$ is in ST.

If $F(z)$ is in ST, then $\int_0^z \frac{F(\zeta)}{\zeta}\, d\zeta$ is in CV.

6. SHARP BOUNDS FOR THE COEFFICIENTS

Although we are still searching for the sharp bound for $|a_n|$ when $f(z)$ is in the parent class S, this problem has been solved for the subclasses ST and CV.

Theorem 6. R. Nevanlinna 1921 [1-B-990]. If

(22) $$f(z) = z + \sum_{k=2}^{\infty} a_k z^k$$

is in ST, then for each positive integer n

(23) $$|a_n| \le n.$$

Further, this inequality is sharp for each index n, and if equality occurs for just one $n \ge 2$, then $f(z)$ is a rotation of the Koebe function.

Proof. If $f(z)$ is in ST, then by Theorems 1 and 4, $zf'(z)/f(z)$ is in P. Hence

(24) $$\frac{zf'(z)}{f(z)} = 1 + \sum_{k=1}^{\infty} b_k z^k \equiv g(z),$$

where $|b_k| \le 2$ for $k = 1, 2, \ldots$. We make a preliminary rotation so that $b_1 \ge 0$. Then by Theorem 7.3 and Problem 7.5, we have $b_1 = 2$ if and only if $g(z) = (1+z)/(1-z)$. From equations (22) and (24) we find that

(25) $$z + \sum_{k=2}^{\infty} ka_k z^k = \left(z + \sum_{k=2}^{\infty} a_k z^k\right)\left(1 + \sum_{k=1}^{\infty} b_k z^k\right).$$

Equating coefficients of z^n in (25) we obtain

(26) $$na_n = a_n + \sum_{k=1}^{n-2} a_{n-k} b_k + b_{n-1}$$

for $n = 3, 4, 5, \ldots$. For $n = 2$ we have

(27) $$2a_2 = a_2 + b_1,$$

and hence $|a_2| = b_1 \le 2$. Further, equality occurs if and only if $g(z) = (1+z)/(1-z)$. Assume that $|a_k| \le k$ for $k = 2, 3, \ldots, n-1$. Then (26) gives

(28) $$|(n-1)a_n| = \left|\sum_{k=1}^{n-2} a_{n-k} b_k + b_{n-1}\right|$$
$$\le \sum_{k=1}^{n-2} 2|a_{n-k}| + 2 \le 2\left(1 + \sum_{k=2}^{n-1} k\right) = (n-1)n.$$

Hence $|a_n| \le n$. Further, if $b_1 < 2$ there will be strict inequality in (28). Therefore $|a_n| < n$ unless $b_2 = 2$ and $g(z) = (1+z)/(1-z)$. But then $a_n = n$ for each $n > 1$ and hence $f(z)$ is the Koebe function. ∎

Using much deeper methods Leung 1978 [3-B-525] proved that if $f(z)$ is in ST then $||a_{n+1}| - |a_n|| \le 1$ for $n = 1, 2, \ldots$. This is much stronger than Theorem 6 and implies that theorem as a corollary.

Theorem 7. C. Loewner 1917 [1-B-880]. If $f(z)$, given by (22), is in CV, then for each positive integer n

$$|a_n| \le 1. \tag{29}$$

Further, this inequality is sharp for each index n, and if equality occurs for just one $n \ge 2$, then $f(z)$ is a rotation of $z/(1-z)$.

Proof. If $f(z)$ is in CV, then by Theorem 5

$$zf'(z) = z + \sum_{n=2}^{\infty} na_n z^n \tag{30}$$

is in ST. By Theorem 6 we have $n|a_n| \le n$. Hence $|a_n| \le 1$. Since $z/(1-z) = z + z^2 + z^3 + \cdots$, the bound $|a_n| \le 1$ is sharp. Further, if $|a_n| = 1$ for just one $n \ge 2$, then $zf'(z)$ is a rotation of the Koebe function. It follows that $f(z)$ is a rotation of $z/(1-z)$. ∎

We will see in Section 9 that the above process can be reversed. Using the Schwarz-Christoffel transformation we can prove that $|a_n| \le 1$ for $f(z)$ in CV. Then the relation $zf'(z) = F(z)$ will give $|a_n| \le n$ for $F(z)$ in ST.

Theorem 8. PLE. Let $f(z)$ be in ST. Then

$$\frac{r}{(1+r)^2} \le |f(z)| \le \frac{r}{(1-r)^2}, \tag{31}$$

$$\frac{1-r}{(1+r)^3} \le |f'(z)| \le \frac{1+r}{(1-r)^3}, \tag{32}$$

and for each $k \ge 2$

$$|f^{(k)}(z)| \le \frac{k!(k+r)}{(1-r)^{k+2}}. \tag{33}$$

All of these inequalities are sharp, with equality if and only if $f(z)$ is some rotation of $k(z)$.

The inequalities (31) and (32) were already proved for the parent class S (Theorems 6.3 and 6.8); and since the Koebe function is also starlike, these inequalities are sharp for the subclass. The upper bound in (33) can be proved by the technique of dominant power series, Theorem 7.5, and the remark that if $f(z)$ is the Koebe function, then

$$f^{(k)}(z) = \frac{k!(k+z)}{(1-z)^{k+2}}, \qquad k = 0, 1, 2, \ldots. \tag{34}$$

For $k = 2$ and 3 in (33) see Problem 6.14. No lower bound in (33) can be expected since $f(z) \equiv z$ is in ST and $f^{(k)}(z) \equiv 0$ for $k = 2, 3, \ldots$.

For a convex function we can expect different bounds.

Theorem 9. Let $f(z)$ be in CV. Then

$$\frac{r}{1+r} \leq |f(z)| \leq \frac{r}{1-r}, \tag{35}$$

$$\frac{1}{(1+r)^2} \leq |f'(z)| \leq \frac{1}{(1-r)^2}, \tag{36}$$

and for each $k \geq 2$,

$$|f^{(k)}(z)| \leq \frac{k!}{(1-r)^{k+1}}. \tag{37}$$

All of these inequalities are sharp, with equality if and only if $f(z)$ is some rotation of $z/(1-z)$.

The bounds (35) and (36) were obtained independently by Gronwall 1916 [1-B-453] and Loewner 1917 [1-B-880].

Proof. All of the upper bounds follow from Loewner's coefficient inequality $|a_n| \leq 1$ (Theorem 7) and the technique of dominant power series.

To prove the left side of (36) we note that if $f(z)$ is convex, then $zf'(z)$ is starlike and hence by (31)

$$\frac{r}{(1+r)^2} \leq |zf'(z)|. \tag{38}$$

Then, the left side of (35) will follow by integrating the inequality $1/(1+r)^2 \leq |f'(z)|$; see Theorem 6.7. ■

7. TWO RADIUS PROBLEMS

It is well known that if $f(z)$ is analytic and $f'(z) \neq 0$ at a certain point z_0, then each circle $|z - z_0| = r$ is mapped by $f(z)$ onto a simple closed curve that is nearly a circle—if r is sufficiently small. To make this soft statement quantitative, we must put some condition on $f(z)$ and be precise about the conclusion that the image curve is "nearly a circle." In this section it is natural to assume that $f(z)$ is in S and to interpret "nearly a circle" as a closed *convex* curve.

Theorem 10. Nevanlinna 1920 [1-B-989][1]. Let $f(z)$ be in S. Then for each $r \leq 2 - \sqrt{3}$, the image of $|z| = r$ is a simple closed convex curve. The number $R_{CV} \equiv 2 - \sqrt{3}$ is sharp. There is a function in S such that for each $r > R_{CV}$ the image of $|z| = r$ is not a convex curve.

Definition 4. The number $R_{CV} = 2 - \sqrt{3}$ is called the *radius of convexity for the set S.*

Before proving this theorem let us put Theorem 10 and Definition 4 in a general setting. We have some set of functions $\mathscr{M}$ and some property $\mathscr{P}$ that $f(z)$ may or may not possess in the disk E_r. In Theorem 10 the set is S and the property $\mathscr{P}$ is that $f(z)$ maps E_r onto a convex region. The problem is to determine $R_{\mathscr{P}}$, the largest number such that every $f(z)$ in $\mathscr{M}$ has the property $\mathscr{P}$ in the disk E_r for each $r < R_{\mathscr{P}}$. This implies that if $r > R_{\mathscr{P}}$, there is at least one $f(z)$ in $\mathscr{M}$ that does not have the property $\mathscr{P}$ in E_r. Formally we have

Definition 5. The number $R_{\mathscr{P}}$ is called the radius for the property $\mathscr{P}$ in the class $\mathscr{M}$ if $R_{\mathscr{P}}$ is the least upper bound of all numbers r such that every $f(z)$ in $\mathscr{M}$ has the property $\mathscr{P}$ in the disk E_r.

This definition gives us the machinery for creating new problems for

[1] Gronwall 1916 [1-B-453] announced this result without proof, but the nature of this paper seems to indicate that his proof was erroneous.

research. Thus if we are given 20 different classes of functions (analytic in the unit disk) and 20 different properties, then the determination of R for each property in each class gives 400 problems. Of course, some of the problems may be meaningless, some may be trivial, and some may be difficult. However, some problems of this type are still open. We will attempt a systematic survey of these "radius problems" in Chapter 13. For the present we concentrate on the prototype, Theorem 10.

Proof of Theorem 10. By Theorem 6.6 (inequality 6.23)

$$1 + \operatorname{Re} \frac{zf''(z)}{f'(z)} \geq 1 + \frac{2r^2 - 4r}{1 - r^2} = \frac{1 - 4r + r^2}{1 - r^2} \tag{39}$$

for every $f(z)$ in S. Hence each $f(z)$ is convex in E_r for every r for which the right side is positive. Since the zeros of $r^2 - 4r + 1$ are $2 \pm \sqrt{3}$, it is clear that if $r < 2 - \sqrt{3}$, then in E_r

$$1 + \operatorname{Re} \frac{zf''(z)}{f'(z)} > 0. \tag{40}$$

Thus $f(z)$ maps E_r onto a convex domain. If $f(z) = z/(1 + z)^2$, then

$$1 + \operatorname{Re} \frac{zf''(z)}{f'(z)} = \operatorname{Re} \left(1 + \frac{-4z + 2z^2}{1 - z^2}\right) = \operatorname{Re} \frac{1 - 4z + z^2}{1 - z^2}$$

and this is negative if $z = r$ and $2 - \sqrt{3} < r < 1$. Thus the radius of convexity[1] in the class S is $2 - \sqrt{3}$. ■

One expects a theorem about R_{ST}, the radius of starlikeness, to accompany Theorem 10 which gives the radius of convexity in the set S. However, there is an essential difficulty. The Koebe function is the extremal function that is convex for $|z| < 2 - \sqrt{3}$, but not in any larger circle. However, this same function is starlike in E and hence cannot be the extremal function for the radius of starlikeness. Inequalities that are sharp for the Koebe function can be useful in proving Theorem 10 but cannot possibly help in determining the radius of starlikeness. Deeper

[1] Equality occurs only if $f(z) = e^{-i\alpha}k(e^{i\alpha}z)$. For an exhaustive study of the problem of "uniqueness" of the solution of an extremal problem, see Campbell 1974 [2-B-212].

methods are needed, and these were supplied by Loewner 1923 [1-B-882]. Here we merely state the result in

Theorem 11. PWO. Grunsky 1933 [1-B-484]. In the set S, the radius of starlikeness is the root of the equation

$$\ln \frac{1+r}{1-r} = \frac{\pi}{2}, \tag{41}$$

namely,

$$R_{ST} = \frac{e^{\pi/2}-1}{e^{\pi/2}+1} = \tanh \frac{\pi}{4} \approx 0.65579. \tag{42}$$

8. INTEGRAL REPRESENTATIONS

In Theorem 7.22 we proved that every normalized function $p(z)$ with positive real part in E has a representation as a Stieltjes integral

$$p(z) = \frac{1}{2\pi}\int_0^{2\pi} \frac{1+ze^{-i\phi}}{1-ze^{-i\phi}}\, d\mu(\phi), \qquad \int_0^{2\pi} d\mu(\phi) = 2\pi, \tag{43}$$

where $\mu(\phi)$ is nondecreasing. Given $p(z)$, $\mu(\phi)$ is not uniquely determined in (43), but it will be uniquely determined by $p(z)$ if we impose the further condition that

$$\mu(\phi) = \frac{1}{2}[\mu(\phi+0)+\mu(\phi-0)].$$

For convenience we make

Definiton 6. Let $M(a, b)$ denote the set of all functions $\mu(\phi)$ that are nondecreasing in the interval $[a, b]$ and satisfy the condition

$$\int_a^b d\mu(\phi) = b-a. \tag{44}$$

Using the integral representation (43) we can derive integral representation for functions in related classes such as P' (see Problem 7.7), CV, and ST.

If we replace $p(z)$ in (43) by $f'(z)$ and integrate, we find that $f(z)$ is in P' if and only if there is a $\mu(\phi)$ in $M(0, 2\pi)$ such that

$$\text{(45)} \qquad f(z) = z - \frac{1}{\pi}\int_0^{2\pi} [z + e^{i\phi} \ln (1 - ze^{-i\phi})]\, d\mu(\phi)$$
$$= -z - \frac{1}{\pi}\int_0^{2\pi} e^{i\phi} \ln (1 - ze^{-i\phi})\, d\mu(\phi).$$

If $f(z)$ is in ST, we replace $p(z)$ by $zf'(z)/f(z)$ in (43), divide by z, and integrate. Here some care is needed because of the first order pole at the origin. The difficulty is avoided if we integrate from z_0 to z where $z_0 \neq 0$ and then let $z_0 \to 0$. We find that $f(z)$ is in ST if and only if

$$\text{(46)} \qquad f(z) = ze^{I(z)},$$

where

$$\text{(47)} \quad I(z) = -\frac{1}{\pi}\int_0^{2\pi} \ln (1 - ze^{-i\phi})\, d\mu(\phi), \qquad \mu(\phi) \text{ in } M(0, 2\pi).$$

Finally, we may replace $p(z)$ by $1 + zf''(z)/f'(z)$ in (43) and integrate to obtain an integral representation for convex functions. Of course, it is simpler to use Alexander's Theorem that $f(z)$ is in CV if and only if $zf'(z)$ is in ST and apply the result obtained in (46) and (47). We find that if $f(z)$ is in CV, then (Robertson 1936 [1-B-1144])

$$\text{(48)} \qquad f(z) = \int_0^z e^{I(\zeta)}\, d\zeta.$$

and conversely.

A formula such as (45), (46), or (48) is often called a *structural formula* for the set of functions that it generates.

Since these integral representations give a complete description of the sets P', ST, and CV, they are very useful in proving theorems about these classes of functions. For example, the coefficient bound $|a_n| \leq n$ for normalized starlike functions follows immediately from formulas (46) and (47). Indeed

$$I(z) = \frac{1}{\pi}\int_0^{2\pi} \left(\sum_{n=1}^{\infty} \frac{1}{n} z^n e^{-in\phi}\right) d\mu(\phi) << 2 \ln \frac{1}{1-z}.$$

Hence

$$f(z) = ze^{I(z)} << ze^{-2 \ln (1-z)} = \frac{z}{(1-z)^2} = \sum_{n=1}^{\infty} nz^n.$$

Nevertheless the information that is implicitly contained in a structural formula is not always easy to retrieve. For example, it is known that there are functions in P' that are *not* starlike (see Zmorovič 1954 [1-B-1519], Gal′perin 1958 [1-B-350], or Krzyz 1962 [3-B-470]). But the precise value of R_{ST} for the set P' is unknown.

9. THE SCHWARZ-CHRISTOFFEL TRANSFORMATION

This transformation is discussed in some detail in most textbooks on function theory (for examples, see Markushevich [1, vol. 2]) so we merely review the facts here.

If $a_1 < a_2 < \cdots < a_m$ and $\mu_1, \mu_2, \ldots, \mu_m$ are real numbers with

$$\sum_{k=1}^{m} \mu_k = 2, \tag{49}$$

then the function

$$f(z) = A \int_0^z \prod_{k=1}^{m} (a_k - \zeta)^{-\mu_k} \, d\zeta + B \tag{50}$$

maps the upper half-plane $\operatorname{Im} z > 0$ onto the interior of a polygon as indicated in Figure 5. The vertices of the polygon are the points $f(a_k)$, and at each vertex the exterior angle is $\mu_k \pi$. Conversely, it can be proved that each simply-connected "unbranched" domain that is bounded by a polygon can be mapped conformally onto the upper half-plane by the inverse of a function that has the form (50) and the exponents satisfy equation (49).

In describing the image domain one is tempted to write that the boundary is a simple closed polygon, but this may be false. As indicated in Figure 6, the polygon may intersect itself, and the simply-connected polygonal domain must be regarded as lying on a Riemann surface. If the domain contains no branch points, it is frequently referred to as a *schlicht* domain. It would be very helpful to have a method for distinguishing the cases in which $f(z)$ given by (50) maps onto a domain of the type shown in Figure 5, rather than a domain of the type shown in Figure 6. More accurately, it would be helpful to have a set of conditions on the constants $a_1, a_2, \ldots, a_m, \mu_1, \mu_2, \ldots, \mu_m$ that are necessary and sufficient for $f(z)$ to be univalent in $\operatorname{Im} z > 0$.

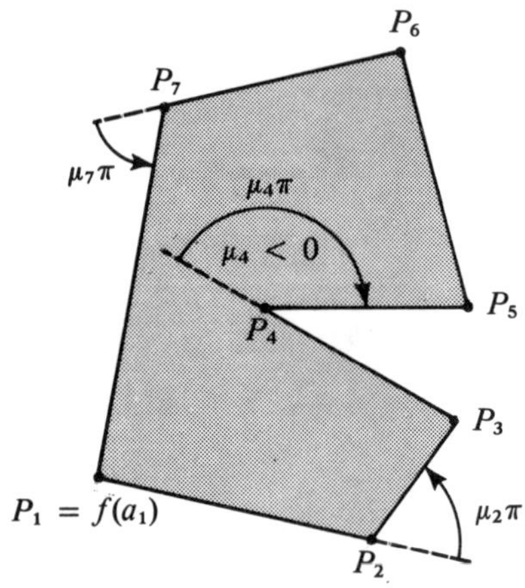

FIGURE 5.

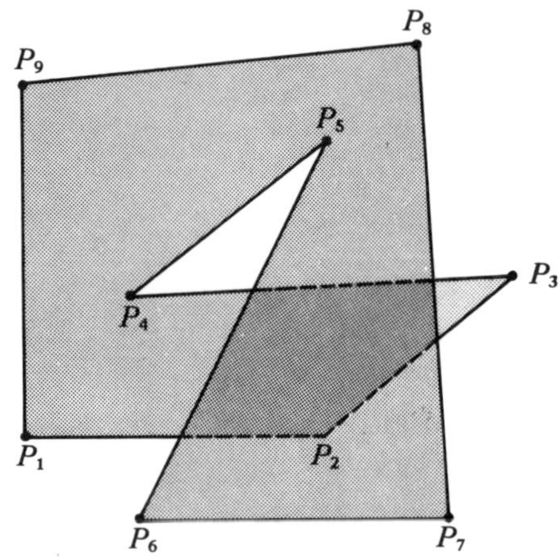

FIGURE 6.

If we map the upper half ζ-plane onto the unit disk, and select the constants A and B so that $f(0) = 0$ and $f'(0) = 1$, then (50) can be put in the form

$$f(z) = \int_0^z \prod_{k=1}^{m} (1 - \zeta e^{-i\phi_k})^{-\mu_k}\, d\zeta. \tag{51}$$

Thus if $\phi_1, \phi_2, \ldots, \phi_m, \mu_1, \mu_2, \ldots, \mu_m$ are real numbers satisfying the condition (49), then the Schwarz-Christoffel transformation (51) maps E onto a polygonal domain with no interior branch points. The point $f(e^{i\phi_k})$ is a vertex of the domain and the exterior angle at that point is $\mu_k\pi$.

The representation (51) is quite useful when $f(E)$ is convex because in this case $\mu_k > 0$ for each k. For $\mu > 0$ we find that

$$\begin{aligned}(1 - \zeta e^{-i\phi})^{-\mu} &\ll (1 - \zeta)^{-\mu} \\ &= 1 + \sum_{n=1}^{\infty} \frac{\mu(\mu+1)\cdots(\mu+n-1)}{n!} z^n.\end{aligned} \tag{52}$$

In (51) each factor is dominated in this manner. Hence

$$f(z) \ll \int_0^z \prod_{k=1}^{m} (1 - \zeta)^{-\mu_k}\, d\zeta \tag{53}$$

$$= \int_0^z \frac{d\zeta}{(1-\zeta)^2} = \frac{z}{1-z} = \sum_{n=1}^{\infty} z^n.$$

Consequently, if $f(z)$ given by (51) maps E onto a convex polygonal domain, then $|a_n| \leq 1$ for each n. Since each convex domain D can be approximated by a sequence of convex polygonal domains that converges to D, this gives a new proof of Theorem 7.

One should note that equation (48) gives equation (51) in the case that D is a convex polygon. Indeed we select for $\mu(\phi)$ a step function that has a jump of $\mu_k\pi$ at ϕ_k for $k = 1, 2, \ldots, m$. Then (47) yields

$$I(z) = -\frac{1}{\pi}\int_0^{2\pi} \ln(1 - ze^{-i\phi})\, d\mu(\phi) = -\frac{1}{\pi}\sum_{k=1}^{m} \mu_k\pi \ln(1 - ze^{-i\phi_k})$$

$$= \ln \prod_{k=1}^{m} (1 - ze^{-i\phi_k})^{-\mu_k}.$$

Consequently, for a convex polygon, the Stieltjes integral representation (48) yields

$$f(z) = \int_0^z e^{I(\zeta)}\, d\zeta = \int_0^z \prod_{k=1}^{m} (1 - \zeta e^{-i\phi_k})^{-\mu_k}\, d\zeta, \tag{54}$$

and this is the Schwarz-Christoffel transformation (51) for the domain. Since a step function for $\mu(\phi)$ in (47) and (48) yields a function that maps E onto a convex polygon, one may expect that in general there is a close relation between $\mu(\phi)$ and the domain $f(E)$ when $f(E)$ is in CV. Similarly, if $f(z)$ is in ST, there should be a close relation between $\mu(\phi)$ in equation (46) and $f(E)$. As Zmorovič 1954 [1-B-1519, p. 177] remarks, "the determination of $\mu(\phi)$ from the shape of $f(E)$ seems to be very difficult and little is known at this time." Some modest contributions in this direction have been made by Yurchenko and L. E. Dunduchenko 1957 [1-B-1501], Keogh 1959 [1-B-665], Pommerenke 1962 [1-B-1073], London and Thomas 1970 [2-B-873], and Sheil-Small 1970 [2-B-1322, p. 591].

At this point we should note that the Schwarz-Christoffel transformation can be generalized to include multisheeted polygons with interior branch points, as both H. A. Schwarz and E. B. Christoffel were aware.

In this generalization we let $\beta_1, \beta_2, \ldots, \beta_{p-1}$ be the critical points in E, where $0 \leq |\beta_k| < 1$ and where any multiple critical point is listed in accordance with its multiplicity. Then in the generalization the integral (51) is replaced by

(55) $$f(z) = C\int_0^z \prod_{k=1}^{m} (1 - \zeta e^{-i\phi_k})^{-\mu_k} \prod_{k=1}^{p-1} (\zeta - \beta_k)(1 - \bar{\beta}_k\zeta)\, d\zeta.$$

Here ϕ_k and μ_k are arbitrary real numbers subject only to the restriction that

(56) $$\sum_{k=1}^{m} \mu_k = 2p, \qquad p \geq 1, p \text{ an integer}.$$

Then $f(E)$ is a multisheeted polygon with m vertices at $f(e^{i\phi_k})$ and $p-1$ branch points at $f(\beta_k)$. Of course in some cases $f(z)$ may have valence greater than p. If the μ_k are all positive, then $f(z)$ is p-valent in E and $f(E)$ is a p-sheeted convex polygon and sharp bounds can be given for the coefficients of the Maclaurin series for functions of this generalized convex class. An Alexander type theorem (see Theorem 5) relates this generalized convex class to a generalized starlike class. For details and further references on this extension, see Goodman 1950 [1-B-431]. For a careful analysis of the various ways of defining starlikeness for a multivalent function, the reader is referred to the excellent work of J. A. Hummel 1967 [2-B-599].

10. ANOTHER LOOK AT THE HERGLOTZ FORMULA

We will try to generalize Theorem 7.22. Let D be some simply-connected domain containing the point A_0 and let $F(z)$ be univalent in E with $F(E) = D$ and $F(0) = A_0$.

Let $SJ(A_0, D)$ be the set of all functions $f(z)$ that can be represented in the form

(57) $$f(z) = \frac{1}{2\pi}\int_0^{2\pi} F(e^{i\phi}z)\, d\mu(\phi)$$

where $\mu(\phi)$ is nondecreasing and

(58) $$\frac{1}{2\pi}\int_0^{2\pi} d\mu(\phi) = 1.$$

If D is the right half-plane and $A_0 = 1$, then $F(z) = (1+z)/(1-z)$, apart from an unimportant rotation, and (57) is just the Herglotz formula for functions in P.

On the other hand, let $AF(A_0, D)$ be the set of all functions $f(z)$

analytic in E with $f(E) \subset D$ and $f(0) = A$. Theorem 7.22 can be reformulated as follows:

If D is the right half-plane and $A_0 = 1$, then

$$SJ(A_0, D) = AF(A_0, D). \tag{59}$$

To generalize Theorem 7.22 we may ask what other domains D have the property that equation (59) holds. The complete answer is given in

Theorem 12. Goodman 1963 [1-B-443]. Equality holds in (59) if and only if D is a half-plane.

Proof. We first prove that $SJ(A_0, D) = AF(A_0, D)$ implies that D is a convex set, and then prove that D is a half-plane. If D is not convex, there is some w_0 not in D, and w_0 lies on some line segment joining two points w_1 and w_2 in D. Thus $w_0 = tw_1 + (1-t)w_2$ for some t in $(0, 1)$. Further, we can select w_1 and w_2 so that they lie on some level curve of $F(z)$. Thus $w_1 = F(re^{i\theta_1})$, and $w_2 = F(re^{i\theta_2})$. By selecting $\mu(\phi)$ so that it has a jump of $2\pi t$ at $\phi = \theta_1$ and a jump of $2\pi(1-t)$ at $\phi = \theta_2$, we obtain an $f(z)$ for which $f(r) = w_0$ and hence $f(E)$ is not contained in D. Thus the condition (59) implies that D is convex.

Next we observe that the relation (59) is invariant under translations, rotations, and similitudes. Thus if $cD + b$ denotes the image of D under $W = cw + b$, where $c \neq 0$, and (59) is true, then (59) remains true for the pair $(cA_0 + b, cD + b)$. These operations permit us to carry any half-plane onto the half-plane $\operatorname{Re} w > 0$ and take a_0 into 1. Thus Herglotz's formula gives the "if" part of Theorem 12.

To prove the converse part, we let (A_0, D) be any pair for which (59) is true. By a translation we can set $A_0 = 0$ and by a shrinking or expansion we can adjust D so that $F'(0) = 1$. Let $F(z) = z + \Sigma_{n=2}^{\infty} A_n z^n$. Since D is convex, Loewner's Theorem gives $|A_n| \leq 1$. Consequently if $f(z) = \Sigma_{n=1}^{\infty} a_n z^n$ is given by (57), then

$$|a_n| = \left| \frac{1}{2\pi} \int_0^{2\pi} A_n e^{in\phi}\, d\mu(\phi) \right| \leq |A_n| \leq 1. \tag{60}$$

On the other hand, for each positive integer n, the function $F(z^n) = z^n + \cdots$ is also in the set $AF(0, D)$. Thus if (59) is true, we can apply the inequality (60) and deduce that for this function $1 = |a_n| \leq A_n \leq 1$. Hence $A_n = 1$ for each n, and D must be the half-plane $\operatorname{Re} w > 1/2$. ∎

PROBLEMS

1. Prove that $L_0(z) \equiv (1+z)/(1-z)$ satisfies the inequality (9) for z in E. Can you prove the same assertion for $L(z) = (az+b)/(cz+d)$?
2. Prove that the Koebe function satisfies the inequality (10) for z in E. Does it satisfy the inequality (7) in E, if $w_0 > 0$?
3. Prove that if $f(E)$ is a convex domain, and Γ is any circle in E (not necessarily with its center at the origin), then $f(\Gamma)$, is a simple closed convex curve. Robertson 1936 [1-B-1144, p. 382]
4. Prove that $f(z) \equiv z + cz^2$ is in CV if and only if $|c| \leq 1/4$. Prove that the function is in ST if and only if $|c| \leq 1/2$.
5. Prove that $-\ln(1-z)$ is in CV.
6. Let $f(z) = z + \Sigma_{n=2}^{\infty} a_n z^n$. Prove that if $\Sigma_{n=2}^{\infty} n|a_n| \leq 1$, then $f(z)$ is in ST. If $\Sigma_{n=2}^{\infty} n^2|a_n| \leq 1$, then $f(z)$ is in CV. Kobori 1932 [1-B-673].
7. Referring to Theorem 3 (Ford's Theorem) let $T(w) = w^2/2$. Sketch a domain D that has the property T for this function, but D is not starlike with respect to the origin.
8. If $f(z)$ is in CV, prove that $f(E)$ will always cover the disk $|w| < 1/2$. For a short proof see MacGregor 1964 [1-B-A-95].
9. Complete the proof that equations (46) and (47) give an integral representation for functions in ST. Observe that equation (48) for functions in CV is a trivial consequence.
10. Prove that if $f(z)$ is in ST, then

$$\ln \sqrt{\frac{z}{f(z)}} = \frac{1}{2\pi}\int_0^{2\pi} \ln(1 - ze^{-i\phi})\, d\mu(\phi),$$

where $\mu(\phi)$ is in $M(0, 2\pi)$.

11. Let

$$f(z) \equiv \frac{1}{2\pi}\int_0^{2\pi} G(ze^{-i\phi})\, d\mu(\phi),$$

where $\mu(\phi)$ is in $M(0, 2\pi)$. Prove that $f(E)$ is contained in the convex hull of $G(E)$. If $G(z)$ is a convex function, then $f(E) \subset G(E)$.

12. Combine the results of Problems 5, 10, and 11 to prove that if $f(z)$ is in ST, then $\sqrt{z/f(z)}$ lies in the disk $|w-1| < 1$. Sharpen this to prove that if $|z| = r$, then

$$|\sqrt{z/f(z)} - 1| \leq r. \tag{61}$$

When can equality occur? The inequality (61) is due to Marx 1932 [1-B-909] (see Strohhäcker 1933 [1-B-1340] and Robertson 1936 [1-B-1144]). The proof outlined in the problems is due to Ašnevic and Ulina 1955 [1-B-60].

13. Prove that if $f(z)$ is in CV, then

$$\left|\frac{1}{\sqrt{f'(z)}} - 1\right| < 1 \quad \text{and} \quad \operatorname{Re}\sqrt{f'(z)} > 1/2. \tag{62}$$

Use this result to find sharp bounds for $\arg f'(z)$.

14. Suppose that λ and μ are nonnegative with $\lambda + \mu = 1$, and $f(z)$ and $g(z)$ are in ST. Prove that fractional powers can be defined so that

$$F(z) \equiv f^{\lambda}(z) g^{\mu}(z) \tag{63}$$

is also in ST. If $f(z)$ and $g(z)$ are in CV, does it follow that $F(z)$ is in CV?

15. Find $F(E)$ if $F(z) = z/\sqrt{1-z^2}$.

16. Let $f(z)$ be the convex function given by equation (51), and suppose that $F(z) = zf'(z)$ is the associated starlike function

$$F(z) = \frac{z}{\prod_{k=1}^{m} (1 - ze^{i\phi_k})^{\mu_k}},$$

where $\mu_k > 0$ and $\Sigma_{k=1}^{m} \mu_k = 2$. Describe $f(E)$.

17. Is $f(z) \equiv \text{Arc tan } z$ in S, ST, or CV? How about $g(z) \equiv \text{Arc sin } z$?

18. Use the result in Problem 7.36 to find the radius of convexity for the set P'. MacGregor 1962 [1-B-887]. The radius of starlikeness in this set is not known.

19. Consider the set of functions $f(z) = z + \cdots$, regular in E (not necessarily univalent in E) and such that $\operatorname{Re}(f(z)/z) > 0$ in E. Find the radius of univalence for this set of functions. Noshiro 1934 [1-B-1000].

20. Under the conditions stated prove that each of the following function is in ST. See Bernardi 1959 [1-B-108].

(a) $$f(z) = z\left(1 + \frac{z}{n}\right)^m, \quad n > 1 + |m|, \ n, m \text{ real.}$$

(b) $$f(z) = ze^{I(z)}, \quad I(z) = \int_0^z b(\zeta)\, d\zeta, \quad |b(\zeta)| < 1, \text{ in } E.$$

In Problems 21 through 28 we study convexity and starlikeness on Γ, a level curve of $g(z)$ where $g(z)$ is analytic and $g'(z) \neq 0$ on the curves under consideration. We consider four different sets of level curves (with $z = x + iy = re^{i\theta}$):

$$\Gamma_1: \ \operatorname{Re} g(z) = u_0, \qquad \Gamma_2: \ \operatorname{Im} g(z) = v_0,$$

$$\Gamma_3: \ |g(z)| = R_0 \qquad \Gamma_4: \ \arg g(z) = \phi_0,$$

where u_0, v_0, R_0, and ϕ_0 are constants. Each of these level curves may be parametrized and a direction is assigned to the curve Γ_k so that:

$$\text{On } \Gamma_1: \ \frac{dv}{dt} > 0, \qquad \text{On } \Gamma_2: \ \frac{du}{dt} > 0,$$

$$\text{On } \Gamma_3: \ \frac{d\phi}{dt} > 0, \qquad \text{On } \Gamma_4: \ \frac{dR}{dt} > 0, \qquad R > 0.$$

See Bernardi 1961 [1-B-109] for more details.

21. Let $\Gamma_k^\star$ be some connected piece of the level curve Γ_k of $g(z)$. Prove that the image of $\Gamma_k^\star$ under $f(z)$ is starlike with respect to w_0 if and only if the following condition is satisfied on $\Gamma_k^\star$:

$$\text{On } \Gamma_1^\star: \ \operatorname{Re} \frac{f'(z)}{(f(z) - w_0)g'(z)} \geq 0,$$

$$\text{On } \Gamma_2^\star: \ \operatorname{Im} \frac{f'(z)}{(f(z) - w_0)g'(z)} \geq 0,$$

$$\text{On } \Gamma_3^\star: \ \operatorname{Re} \frac{f'(z)g(z)}{(f(z) - w_0)g'(z)} \geq 0,$$

$$\text{On } \Gamma_4^\star: \ \operatorname{Im} \frac{f'(z)g(z)}{(f(z) - w_0)g'(z)} \geq 0.$$

These four inequalities may be derived independently. Observe that the second condition can be derived from the first condition by setting $g_2(z) \equiv -ig(z)$, and the last two follow from the first two by setting $G(z) = \ln g(z)$.

22. Derive the following inequalities for the level arcs of $g(z)$ to be starlike with respect to z_0:

$$\text{On } \Gamma_1^\star: \ \operatorname{Re} (z - z_0)g'(z) \geq 0, \qquad \text{On } \Gamma_2^\star: \ \operatorname{Im} (z - z_0)g'(z) \leq 0,$$

$$\text{On } \Gamma_3^\star: \ \operatorname{Re} (z - z_0)\frac{g'(z)}{g(z)} \geq 0, \qquad \text{On } \Gamma_4^\star: \ \operatorname{Im} (z - z_0)\frac{g'(z)}{g(z)} \leq 0.$$

23. Find a domain D and a function $g(z)$ such that all four or the level arcs $\Gamma_k^\star$ in D are starlike with respect to the origin.

24. With the notation introduced above prove that the image of $\Gamma_k^\star$ under $f(z)$ is convex if and only if the following condition is satisfied on $\Gamma_k^\star$:

$$\text{On } \Gamma_1^\star:\ \operatorname{Re}\left[\frac{f''(z)}{f'(z)g'(z)} - \frac{g''(z)}{(g'(z))^2}\right] \geq 0.$$

$$\text{On } \Gamma_2^\star:\ \operatorname{Im}\left[\frac{f''(z)}{f'(z)g'(z)} - \frac{g''(z)}{(g'(z))^2}\right] \geq 0,$$

$$\text{On } \Gamma_3^\star:\ \operatorname{Re}\left[1 + \frac{f''(z)g(z)}{f'(z)g'(z)} - \frac{g(z)g''(z)}{(g'(z))^2}\right] \geq 0,$$

$$\text{On } \Gamma_4^\star:\ \operatorname{Im}\left[\frac{f''(z)g(z)}{f'(z)g'(z)} - \frac{g(z)g''(z)}{(g'(z))^2}\right] \geq 0.$$

25. Derive the following inequalities for the level arcs of $g(z)$ to be convex:

$$\text{On } \Gamma_1^\star:\ \operatorname{Re}\frac{g''(z)}{(g'(z))^2} \leq 0, \qquad \text{On } \Gamma_2^\star:\ \operatorname{Im}\frac{g''(z)}{(g'(z))^2} \leq 0,$$

$$\text{On } \Gamma_3^\star:\ \operatorname{Re}\frac{g(z)g''(z)}{(g'(z))^2} \leq 1, \qquad \text{On } \Gamma_4^\star:\ \operatorname{Im}\frac{g(z)g''(z)}{(g'(z))^2} \leq 0.$$

26. Find a domain D and a function $g(z)$ such that all four of the level arcs $\Gamma_k^\star$ in D are convex.

27. Check the formulas in Problems 21 and 24 by setting $g(z) = z$ to obtain the formulas in Lemmas 1 and 2.

28. Is there an Alexander type theorem (see Theorem 5) relating the convexity and starlikeness conditions in Problems 21 and 24?

29. Find all functions that are regular in E and map each radial line in E onto a convex arc.

In Problems 30 through 36 let $CV^{(k)}$ and $ST^{(k)}$ be the sets of normalized univalent functions that are convex and starlike, respectively, in E and in addition have k-fold symmetry (see Definition 2.3 and Theorem 2.3). In these problems $k \geq 2$ is an integer.

30. If $f(z)$ is in ST, is it true that $g(z) \equiv f(z^k)^{1/k}$ is always in $ST^{(k)}$? Here, of course, we select that branch of the kth root for which $g'(0) = 1$.

31. If $f(z)$ is in CV, is it true that $g(z) \equiv f(z^k)^{1/k}$ is always in $CV^{(k)}$?

32. Sharp bounds for $|f(z)|$ and $|f'(z)|$ in the set $S^{(k)}$ were found in Problems 6.10 and 6.12. Since the extremal function for these bounds is starlike, these bounds are also the sharp bounds in the set $ST^{(k)}$. Find sharp bounds for these two quantities in the set $CV^{(k)}$. Gronwall 1916 [1-B-454].

33. Prove that functions in $ST^{(k)}$ and $CV^{(k)}$ have the following integral representations:

$$ST^{(k)}: f(z) = ze^{J(z)}, \qquad CV^{(k)}: f(z) = \int_0^z e^{J(\zeta)}\, d\zeta$$

where

$$J(z) = -\frac{1}{k\pi}\int_0^{2\pi} \ln(1 - z^k e^{-i\phi})\, d\mu(\phi), \qquad \mu(\phi) \text{ in } M(0, 2\pi).$$

Robertson 1936 [1-B-1144].

34. Find sharp bounds for the coefficients in

$$f(z) = z + \sum_{n=1}^{\infty} b_{nk+1} z^{nk+1} \tag{64}$$

if: (I) $f(z)$ is in $ST^{(k)}$, and (II) $f(z)$ is in $CV^{(k)}$, Goluzin 1929 [1-B-381, p. 170], Levin 1935 [1-B-846], and Robertson 1936 [1-B-1144, p. 380]. For some generalizations see Poole 1966 [2-B-1170] and MacGregor 1963 [1-B-888]. For $k = 2$, see Privalov 1924 [2-B-1083].

35. Assuming that the radius of starlikeness $R_{ST} = \tanh(\pi/4)$ for the class S (see Theorem 11) find R_{ST} for the class of $S^{(k)}$.

36. Show that the function $f(z) \equiv z/(1 - z^k)^{2/k}$ is convex in the disk

$$|z| < \sqrt[k]{k+1-\sqrt{k^2+2k}} \equiv R_{CV}, \tag{65}$$

and not in any larger disk. The proof that R_{CV} is the radius of convexity in $S^{(k)}$ is somewhat difficult. Goluzin 1935 [1-B-383, p. 171], who uses an inequality of Loewner 1919 [1-B-881]. See Pflanz 1935 [1-B-1043, p. 81].

37. Give a geometric argument to prove that if $0 \le \alpha < \beta < 2\pi$ and $f(z)$ is in CV, then

$$F(z) \equiv \frac{f(e^{i\beta}z) - f(e^{i\alpha}z)}{e^{i\beta} - e^{i\alpha}} \tag{66}$$

is in ST. Nehari 1939 [1-B-963] and Rahmanov 1951 [1-B-1090].

38. Does the result stated in Problem 37 have an inverse? Given $F(z)$ in ST and fixed α, β is there always a unique $f(z)$ that satisfies equation (66)? If there is such an $f(z)$, will it be in CV? Will it be univalent?

39. Show that one part of Alexander's Theorem is a special case of the result in Problem 37.

★ ★ **40.** Prove that if $f(z)$ is convex, then for each z_0 in E, the function

$$F(z) = z\left(\frac{f(z)-f(z_0)}{z-z_0}\right)^2 \tag{67}$$

is starlike in E. Sheil-Small 1969 [2-B-1319]. Maria Fait and Zlotkiewicz 1976 [3-B-239] have proved that if $0 \leq \alpha \leq 2$ and $f(z)$ is in CV, then $F(z) \equiv z\left(\frac{f(z)-f(z_0)}{z-z_0}\right)^\alpha$ is in $ST(1-\alpha/2)$; see Chapter 9.1.

41. Use the result in Problem 40 with $z_0 = 0$ to prove that if $f_0(z)$ is in CV, then the transformation

$$f_{n+1}(z) = \int_0^z \left(\frac{f_n(\zeta)}{\zeta}\right)^2 d\zeta, \qquad n = 0, 1, 2, \ldots \tag{68}$$

generates an infinite sequence of functions, each in CV. Find $f_1(z)$ if $f_0(z) = z + z^2/4$.

42. Find all normalized convex functions that are invariant under the transformation (68).

43. Let $K(\Gamma) \equiv K(\Gamma(z))$ denote the curvature of a directed curve at a point z, and let $K(f(\Gamma))$ denote the curvature of the image of Γ under $f(z)$ at the image point. If $e^{i\alpha}$ specifies the direction of the tangent to Γ at z, prove that

$$K(f(\Gamma)) = \frac{1}{|f'(z)|}\left(K(\Gamma) + \operatorname{Im}\left(e^{i\alpha}\frac{f''(z)}{f'(z)}\right)\right). \tag{69}$$

Study 1913 [1-B-1341].

44. Use the result in Problem 43 to show that if Γ is C_r, the counterclockwise circle $|z| = r$, then

$$K(f(C_r)) = \frac{1}{|zf'(z)|}\operatorname{Re}\left(1 + \frac{zf''(z)}{f'(z)}\right).$$

Note that this result also gives the condition that $f(C_r)$ is a convex curve.

★ **45.** Prove that if $f(z)$ is in ST, then for each z in E and each directed curve Γ through z,

$$|K(f(\Gamma))| \leq \frac{(1+r)^3}{1-r}\,|K(\Gamma)| + (4-2r)\left(\frac{1+r}{1-r}\right)^2,$$

and this inequality is sharp. Equality occurs for a suitable rotation of the Koebe function. See Goodman 1968 [2-B-474]. Hint: If $\operatorname{Re} zf'(z)/f(z) \geq 0$ in E, the same is true for the reciprocal $f(z)/zf'(z)$.

★ **46.** Prove that if $f(z)$ is in CV then for each circle $|z| = r$

(70)
$$K(f(C_r)) \geq \frac{1-r^2}{r}.$$

Hint: It may be simpler to find an upper bound for the radius of curvature. Zmorovič 1952 [1-B-1517] and Eenigenburg 1970 [2-B-350] have also investigated those functions for which equality occurs in (70). Since the extremal function is not unique, this appears to be a more difficult problem. Zmorovič [1-B-1517] also found the sharp upper bound for $K(f(C_r))$ in the set CV and the sharp upper and lower bounds for K in the set $CV^{(k)}$. Eenigenburg [2-B-350] found the sharp lower bound for K in the set $CV(\alpha)$; see Chapter 9.1. For further work on problems of curvature see Mirošničenko [2-B-972 and 973] and Korickii [1-B-731 and 732].

47. For fixed ζ in E set

(71)
$$g(z) = -\left[f\left(\frac{-z+\zeta}{1-\bar{\zeta}z}\right) - f(\zeta)\right]\frac{1}{f'(\zeta)(1-|\zeta|^2)}.$$

Prove that if $f(z)$ is in S, then $g(z)$ is in S, and that

(72)
$$\frac{\zeta}{g(\zeta)} = (1-|\zeta|^2)\,\frac{\zeta f'(\zeta)}{f(\zeta)}.$$

Conversely, prove that if $g(z)$ is in S, then for each ζ in E there is an $f(z)$ in S for which (72) holds. See Marx 1932 [1-B-909, p. 49].

48. Let $R_\alpha(q(z))$ be the largest R with the property that if $f(z)$ is in S, then $|\arg q(z)| < \alpha$ for $|z| < R$. Use the result of Marx from Problem 47 to show that $R_\alpha(q(z))$ is the same for $q_1(z) \equiv zf'(z)/f(z)$ and $q_2(z) \equiv f(z)/z$.

49. Prove that R_{ST}, the radius of starlikeness for the set S, is the largest number R such that Re $(f(z)/z) > 0$ for $|z| < R$ and every $f(z)$ in S.

★ **50.** From Problem 2.32 the function $f(z) = (z - cz^2)/(1 - z)^2$ is in S if and only if $|2c - 1| \leq 1$. Use this and the results in Problems 47, 48, and 49 to prove that $R_{ST} \leq 4\sqrt{2} - 5 \approx 0.65686$, Marx [1-B-909, p. 56]. It is remarkable that he came so close to the true value $R_{ST} = \tanh(\pi/4) \approx 0.65579$ using only a rational function. Hint: With a proper selection of α in $c = (1 + e^{i\alpha})/2$ show that Re $(z/f(z)) > 0$ if and only if Re $(1 - z)^2(2 - \bar{z}) > |z(1 - z)^2|$. If $z = x + iy$, this reduces to $\phi(r) \equiv 2 - r - r^3 - (5 - 2r + r^2)x + 4x^2 > 0$. Select x so that $\phi(r)$ is a minimum and find a root of $\phi(r) = 0$.

51. Prove that if $g(z)$ is in $ST^{(2)}$, then Re $(g(z)/z) > 1/2$ in E. Is this bound sharp? Hint: Use Problem 12.

★ **52.** Let $f(z)$ be a normalized polynomial with roots at $0, z_1, z_2, \ldots, z_{n-1}$. Set $g(z) = z(z - |z_1|)(z - |z_2|) \cdots (z - |z_{n-1}|)$, and let R be the smallest zero of $g'(z)$. Prove that $f(z)$ is starlike univalent in $|z| < R$. See Obrechkoff 1934 [1-B-1001 and 1004].

★ **53.** The coefficients of the Koebe function satisfy the inequality $|2a_n - a_2a_{n-1}| \leq 2$. This suggests that this might be true for all $f(z)$ in S. Prove that such a conjecture is false even in the smaller set ST. Hint: Consider $f(z) = k(z)(1 - z)^\alpha/(1 + z)^\alpha$ where $0 < \alpha < 2$ and $\alpha \neq 1$. Then $2 < 2a_5 - a_2a_4 < 7/3$. Barnard 1975 [2-B-113].

54. A sequence $b_1, b_2, \ldots, b_n, \ldots$ is called a *subordinating factor sequence* for the set $\mathcal{M}$ if $f(z) = \Sigma_1^\infty a_nz^n$ is in $\mathcal{M}$ implies that $g(z) \equiv \Sigma_1^\infty a_nb_nz^n \prec f(z)$ in E. Prove that $\{b_n\}_1^\infty$ is a subordinating factor sequence for the set CV if and only if $1 + 2\Sigma_1^\infty b_nz^n$ is in P. Hint: Consider the special function $z/(1 - z)$. For the converse use equation (7.19) for b_n. Wilf 1961 [1-B-1477] for Problems 54 through 56. For further work on these sequences see Bernardi 1966 [2-B-139].

55. If $\{b_n\}_1^\infty$ and $\{c_n\}_1^\infty$ are subordinating factor sequences for the set CV, does it follow that $\{b_nc_n\}_1^\infty$ is also a subordinating factor sequence for CV?

56. Prove that if the two functions

$$f(z) = 1 + 2\sum_{n=1}^{\infty} b_nz^n, \qquad \text{and} \qquad g(z) = 1 + 2\sum_{n=1}^{\infty} c_nz^n$$

are in P, then the Hadamard product (see Chapter 14.1)

$$H(z) \equiv 1 + 2 \sum_{1}^{\infty} b_n c_n z^n$$

is also in P. See Nehari and Netanyahu 1957 [1-B-985] and Komatu 1958 [1-B-719].

57. Suppose that $f(z)$ is regular in E and let z_0 be a point on C_r at which $|f(z)|$ has a relative extreme on C_r. Assume that $f(z_0) \neq 0$ and $f'(z_0) \neq 0$. Prove that $Q_{ST}(z_0) \equiv z_0 f'(z_0)/f(z_0)$ is real. See equation (9.1).

58. Show by examples that at a maximum point of $|f(z)|$ on C_r, $Q_{ST}(z_0)$ may be any positive number.

59. Prove the following theorem. Suppose that $f(z)$ is regular in $\bar{E}_r$, $f(0) = 0$, and $|f(z)|$ assumes its maximum at z_0 on C_r. Then $z_0 f'(z_0)/f(z_0) \geq 1$. Hint: Normalize by setting $z_0 = 1$ and let $B(z) = f(rz)/M$. Then use Problem 7.41. This theorem is often called Jack's Lemma, see Jack 1971 [2-B-619].

60. Suppose that $f(z)$ satisfies the conditions of Problem 59 and in addition $f^{(k)}(0) = 0$ for $0 \leq k \leq n$. Prove that $Q_{ST}(z_0) \geq n+1$. Hallenbeck and Ruscheweyh 1975 [2-B-552], and S. Miller and Mocanu 1978 [3-B-595].

61. Prove a theorem, similar to Alexander's Theorem, that relates the two sets P' and ST. See Gal′perin 1958 [1-B-350].

CHAPTER 9

MORE ABOUT STARLIKE AND CONVEX FUNCTIONS

A complete account of the research on convex and starlike functions could easily occupy a book. In this chapter we select a few items from this vast supply of material.

1. CONVEX AND STARLIKE FUNCTIONS OF ORDER ALPHA

For brevity we introduce the notations

$$Q_{ST}(z) \equiv z\frac{f'(z)}{f(z)} \tag{1}$$

and

$$Q_{CV}(z) \equiv 1 + \frac{zf''(z)}{f'(z)}. \tag{2}$$

We recall that with the usual normalization

$$f(z) = z + \sum_{n=2}^{\infty} a_n z^n \tag{3}$$

the condition

$$Re\ Q_{ST}(z) = \frac{d}{d\theta} \arg f(z) > 0 \tag{4}$$

is necessary and sufficient for $f(z)$ to be starlike in E. Further (see Figure 8.3),

$$\text{Re}\ Q_{CV}(z) = \frac{d}{d\theta} \arg izf'(z) = \frac{d\tau}{d\theta} > 0 \tag{5}$$

(where τ is the angle the tangent vector makes with the positive real axis) is necessary and sufficient for $f(z)$ to be convex in E.

One can alter the conditions (4) and (5) by setting other limitations on the behavior of $\arg f(z)$ and τ. M. S. Robertson 1936 [1-B-1144] introduced the concepts of functions starlike and convex of order α.

Definition 1. A function $f(z)$ of the form (3) is said to be *starlike of order* α in E if for all z in E

$$\text{Re}\ Q_{ST}(z) \geq \alpha. \tag{6}$$

The set of all such functions is denoted by $ST(\alpha)$. A function $f(z)$ of the form (3) is said to be *convex of order* α in E if for all z in E

$$\text{Re}\ Q_{CV}(z) \geq \alpha. \tag{7}$$

The set of all such functions is denoted by $CV(\alpha)$.

Since $Q_{ST}(0) = Q_{CV}(0) = 1$, it is clear that we must have $\alpha \leq 1$, otherwise the two sets $ST(\alpha)$ and $CV(\alpha)$ are empty. Further, if $\alpha = 1$, then the sets $ST(\alpha)$ and $CV(\alpha)$ have only one function: $f(z) \equiv z$. If $\alpha < 0$, then a function in either of these sets may fail to be univalent. We will explore these possibilities further in the problems. For the most part we impose the natural condition $0 \leq \alpha \leq 1$.

One can alter the condition $\text{Re}\ Q > 0$ in a different, but equally interesting way, by demanding that $|\arg Q(z)| < \alpha\pi/2$.

Definition 2. Brannan and Kirwan 1969 [2-B-179], Stankiewicz

1966 [2-B-1383] and 1969 [2-B-1386]. A function $f(z)$ of the form (3) is said to be *strongly starlike of order* α in E if for all z in E

$$|\arg Q_{ST}(z)| \leq \alpha\frac{\pi}{2}. \tag{8}$$

The set of all such functions is denoted by $\widetilde{ST}(\alpha)$. A function of the form (3) is said to be *strongly convex of order* α in E if for all z in E

$$|\arg Q_{CV}(z)| \leq \alpha\frac{\pi}{2}. \tag{9}$$

The set of all such functions is denoted by $\widetilde{CV}(\alpha)$.

Again, it is natural to require that $0 \leq \alpha \leq 1$, but the sets $\widetilde{ST}(\alpha)$ and $\widetilde{CV}(\alpha)$ with $\alpha > 1$ may have interesting properties.

We should observe that as α increases, the sets $ST(\alpha)$ and $CV(\alpha)$ become smaller. However, as α increases, the sets $\widetilde{ST}(\alpha)$ and $\widetilde{CV}(\alpha)$ become larger

Theorem 1. PLE. Robertson 1936 [1-B-1144]. Let $f(z)$ be in $CV(\alpha)$ where $0 \leq \alpha \leq 1$. Then

$$\frac{1}{(1+r)^{2(1-\alpha)}} \leq |f'(re^{i\theta})| \leq \frac{1}{(1-r)^{2(1-\alpha)}}. \tag{10}$$

If $\alpha \neq 1/2$, then

$$\frac{(1+r)^{2\alpha-1}-1}{2\alpha-1} \leq |f(re^{i\theta})| \leq \frac{1-(1-r)^{2\alpha-1}}{2\alpha-1}, \tag{11}$$

and if $\alpha = 1/2$, then

$$\ln(1+r) \leq |f(re^{i\theta})| \leq -\ln(1-r). \tag{12}$$

All of these inequalities are sharp. The extremal functions are rotations of

$$f(z) = \frac{1-(1-z)^{2\alpha-1}}{2\alpha-1}; \qquad \text{if } \alpha \neq 1/2, \tag{13a}$$

$$f(z) = -\ln(1-z), \qquad \text{if } \alpha = 1/2. \tag{13b}$$

For a proof, we observe that if $\operatorname{Re} Q_{CV}(z) \geq \alpha$ and $0 \leq \alpha < 1$, then

(14) $$\frac{1}{1-\alpha}\left\{1+z\frac{f''(z)}{f'(z)}-\alpha\right\} = 1+b_1z+b_2z^2+\cdots$$

is in P, and known results for functions in P can be applied.

Obviously, an Alexander type theorem relates these new classes. Assuming that $f(z)$ and $F(z)$ have the normalization (3), we find:

(15a) $f(z)$ is in $CV(\alpha)$ if and only if $F(z) \equiv zf'(z)$ is in $ST(\alpha)$

and

(15b) $f(z)$ is in $\widetilde{CV}(\alpha)$ if and only if $F(z) \equiv zf'(z)$ is in $\widetilde{ST}(\alpha)$.

As a result, Theorem 1 gives sharp bounds for $|F(z)|$ when $F(z)$ is in $ST(\alpha)$, namely

(16) $$\frac{r}{(1+r)^{2(1-\alpha)}} \le |F(re^{i\theta})| \le \frac{r}{(1-r)^{2(1-\alpha)}}.$$

Here an extremal function is

(17) $$k^*(z,\alpha) \equiv \frac{z}{(1-z)^{2(1-\alpha)}}.$$

The same methods applied to the sets $\widetilde{ST}(\alpha)$ and $\widetilde{CV}(\alpha)$ will lead to similar sharp bounds, but the formulas for these bounds are rather complicated. If $F(z)$ is in $\widetilde{ST}(\alpha)$, we can write

$$\frac{zF'(z)}{F(z)} = p^{\alpha}(z)$$

where $p(z)$ is in P. For details see Brannan and Kirwan 1969 [2-B-179].

Theorem 2. PLE. Robertson 1936 [1-B-1144]. Suppose that $F(z) = z+\Sigma_2^\infty A_nz^n$ is in $ST(\alpha)$ with $0 \le \alpha \le 1$. Then for $n = 2, 3, \ldots$

(18) $$|A_n| \le \frac{1}{(n-1)!}\prod_{k=2}^{n}(k-2\alpha).$$

If $f(z)$ given by (3) is in $CV(\alpha)$, then

(19) $$|a_n| \le \frac{1}{n!}\prod_{k=2}^{n}(k-2\alpha).$$

Both inequalities are sharp for each integer $n \geq 2$. An extremal function for (18) is $z/(1-z)^{2(1-\alpha)}$. For (19) the extremal functions are given in equations (13a) and (13b).

Sharp coefficient bounds for the sets $\widetilde{ST}(\alpha)$ and $\widetilde{CV}(\alpha)$ are much harder to obtain, and only partial results are known. See Brannan, Clunie, and Kirwan 1970 [2-B-176].

Other subsets of ST may be interesting. Suppose that α and β are fixed with $-2 < \beta < 0 < \alpha < 2$. Bucka and Ciozda 1973 [2-B-191] studied the set of normalized functions for which

$$\beta\frac{\pi}{2} < \arg Q_{ST}(z) < \alpha\frac{\pi}{2}.$$

In [2-B-192] the same authors examined the subset of ST for which

$$|\arg (Q_{ST}(z) - \alpha)| < \gamma\frac{\pi}{2},$$

where $0 \leq \alpha < 1$ and $0 < \gamma \leq 1$. Using subordination, $f(z)$ is in this set if and only if

$$Q_{ST}(z) \prec (1-\alpha)\left(\frac{1+z}{1-z}\right)^{\gamma} + \alpha.$$

For each natural number k set

$$f^{(k)}(z) = \sum_{n=0}^{\infty} b_n^{(k)} z^n = \sum_{n=0}^{\infty} \frac{f^{(n+k)}(0)}{n!} z^n.$$

The Hurwitz condition

$$\textbf{(20)} \qquad \sum_{n=2}^{\infty} n|b_n^{(k)}| \leq |b_1^{(k)}|$$

assures us that $f^{(k)}(z)$ is starlike univalent in E (or $f^{(k)}(z) \equiv 0$), see Problems 2.7 and 8.6. Buckholtz and S. M. Shah 1980 [3-B-147] consider the subset of ST of all functions that satisfy the inequality (20) for every $k > 1$. They also consider the analogous subset of CV where n is replaced by n^2 in (20) so that each derivative is convex in E.

For further work on the sets $CV(\alpha)$ and $ST(\alpha)$ see Schild 1965 [1-B-A-138], Pinchuk 1968 [2-B-1122], Klein 1968 [2-B-712], and Zawadzki 1971 [2-B-1537].

Using variational methods Pinchuk obtained rotation theorems for

$CV(\alpha)$ and $ST(\alpha)$. If $f(z)$ is in $CV(\alpha)$, then $|\arg f'(z)| \leq 2(1-\alpha)$ Arc sin r. The bound for $|\arg f'(z)|$ in $ST(\alpha)$ is more complicated.

2. ALPHA-CONVEX FUNCTIONS

In 1969 P. Mocanu [2-B-990] introduced the concept of an α-convex function.

Definition 3. A function $f(z)$ of the form (3) is said to be *α-convex in E* (or briefly α-convex) if it is regular,

$$\frac{f(z)}{z} f'(z) \neq 0, \tag{21}$$

and

$$\operatorname{Re}\left[\alpha\left(1 + z\frac{f''(z)}{f'(z)}\right) + (1-\alpha)z\frac{f'(z)}{f(z)}\right] \geq 0 \tag{22}$$

for all z in E. The set of all such functions is denoted by $\alpha - CV$.

The definition is meaningful if α is a complex number, but henceforth we will assume that α is real. The term α-convex is used because α multiplies the expression Q_{CV} which enters in the condition for $f(z)$ to be convex. Some authors use the term α-starlike for the functions that satisfy Definition 3, for example Sakaguchi and Fukui 1979 [3-B-782].

Of course, if $\alpha = 1$, then an α-convex function is convex; and if $\alpha = 0$, an α-convex function is starlike.Thus the sets $\alpha - CV$ give a "continuous" passage from convex functions to starlike functions. However, it is not necessary to restrict α to the interval [0, 1] as we will soon see.

Mocanu [2-B-990] gave a geometric interpretation for the inequality (22), and in fact he started with a geometric condition and derived (22) as a consequence.

One very surprising result is contained in

Theorem 3. Let α be an arbitrary real number, and suppose that $f(z)$ is α-convex. If $\alpha \geq 1$, then $f(z)$ is convex. If $\alpha < 1$, then $f(z)$ is starlike.

Theorem 3, in this form, was first announced and proved by S. Miller, Mocanu, and Reade in 1973 [2-B-968]. However, it should be observed that Sakaguchi in 1962 [1-B-1220] obtained essentially the same result much earlier and prior to the introduction of the term α-convex. Sakaguchi considers more generally functions of the form

$$f(z) = z^p + \sum_{n=p+1}^{\infty} a_n z^n \tag{23}$$

where p is a positive integer, and he imposes the condition

$$\operatorname{Re}\left[1 + \frac{zf''(z)}{f'(z)} + kz\frac{f'(z)}{f(z)}\right] \geq 0 \tag{24}$$

for z in E. He proved that if $k = -1$, there is only one function that satisfies (24), namely $f(z) \equiv z^p$. If $-1 < k \leq 0$, then $f(z)$ is p-valent convex; and if $0 < k$, then $f(z)$ is p-valent starlike. We can pass from (24) back to (22) if we divide by $1 + k > 0$ and set $\alpha = 1/(1+k)$. The range $k > -1$ gives $0 < \alpha < \infty$, so Sakaguchi's results give only half of Theorem 3. However, his argument can be extended to obtain the other half of the theorem.

In the literature the class $\alpha - CV$ is often denoted by $M(\alpha)$ in honor of Mocanu. The notation $MS(\alpha)$ might be a suitable alternative notation for the class $\alpha - CV$.

Proof of Theorem 3. A computation gives

$$\begin{aligned}\frac{\partial}{\partial\theta} \arg Q_{ST}(z) &= \operatorname{Re}\left(1 + \frac{zf''(z)}{f'(z)} - \frac{zf'(z)}{f(z)}\right) \\ &= \operatorname{Re}\left(Q_{CV}(z) - Q_{ST}(z)\right)\end{aligned} \tag{25}$$

wherever $Q_{ST}(z)$ is regular. But $Q_{ST}(0) = 1$ and hence in a small neighborhood of the origin each circle goes into a curve Γ_r contained in $\operatorname{Re} w > 0$ under $w = Q_{ST}(z)$. Let $\varrho > 0$ be the smallest value of r such that Γ_ρ meets the imaginary axis, and since ϱ has this minimum property, the curve will be tangent to the imaginary axis at some point $w_0 = Q_{ST}(z_0)$ where $|z_0| = \varrho$. Since $f'(z) \neq 0$, we have $w_0 \neq 0$, and consequently at z_0 the tangency property gives

$$\frac{\partial}{\partial\theta} \arg Q_{ST}(z_0) = 0 \tag{26}$$

and at the same point

(27) $$\text{Re } Q_{ST}(z_0) = 0.$$

If we conbine (25), (26), and (27), we see that Re $Q_{CV}(z_0) = 0$. Hence for any real α

(28) $$\text{Re } [H(z_0)] \equiv \text{Re } [\alpha Q_{CV}(z_0) + (1-\alpha)Q_{ST}(z_0)] = 0.$$

But Re $[H(z)]$ is a harmonic function and by hypothesis (equation (22)) is nonnegative in E. If z_0 is an interior point of E, then Re $H(z) \equiv 0$ in E. But this is an impossibility, because Re $[H(0)] = \alpha + (1-\alpha) = 1$. Thus the point z_0 (if there is such a point with the property described) must be outside E. Then Re $Q_{ST}(z) > 0$ in E and hence $f(z)$ is starlike in E for all real α.

Suppose further, that $\alpha > 1$. Then from (22) we have that in E

(29) $$\text{Re } [\alpha Q_{CV}(z)] \geq (\alpha - 1) \text{ Re } Q_{ST}(z) > 0$$

from the first half of the proof. Hence $f(z)$ is convex in E. ■

Definition 3 can be simplified by dropping the condition (21). As Sakaguchi and Fukui 1979 [3-B-782] prove, if $f(z)$ satisfies the inequality (22), then $f(z)f'(z) \neq 0$ for $0 < |z| < 1$.

Since the set $0 - CV$ coincides with ST and the set $1 - CV$ is the set CV, we may expect that as α increases from 0 to ∞, the sets $\alpha - CV$ decrease and will be nested.

Theorem 4. Sakaguchi 1962 [1-B-1220]. If $f(z)$ is α-convex and $0 \leq \beta < \alpha$, then $f(z)$ is β-convex. Thus

(30) $$\alpha - CV \subset \beta - CV.$$

Proof. We write the condition (22) in the form

(31) $$\text{Re}\left(\frac{zf'(z)}{f(z)}\right) + \alpha \text{ Re}\left(1 + \frac{zf''(z)}{f'(z)} - \frac{zf'(z)}{f(z)}\right) \geq 0,$$

or more briefly as

(32) $$B + \alpha M \geq 0$$

so that (32) appears as the equation of a straight line in α for each fixed z_0. Now $B \equiv \text{Re } (zf'(z)/f(z)) > 0$ by the proof of Theorem 3. If (32) is true for some $\alpha > 0$, then it is true for every β with $0 \le \beta < \alpha$. ■

If α and β are negative, the roles are reversed.

Theorem 5. S. Miller, Mocanu, and Reade 1973 [2-B-968]. If $f(z)$ is in $\alpha - CV$ and $\alpha < \beta \le 0$, then $f(z)$ is in $\beta - CV$. Thus, we again have equation (30).

The proof of Theorem 5 is the same as that of Theorem 4.

Theorems 4 and 5 tell us that as α grows from $-\infty$ to $+\infty$ the sets $\alpha - CV$ grow steadily to a maximum for $\alpha = 0$ and then decrease steadily. The reader should observe that $f(z) \equiv z$ is in $\alpha - CV$ for every α.

Suppose now that $f(z)$ is α-convex. We set

$$F(z) \equiv f(z)\left[\frac{zf'(z)}{f(z)}\right]^{\alpha}, \qquad F(0) = 0, \;\; F'(0) = 1, \tag{33}$$

where, if α is not an integer, we can select a suitable branch so that $F(z)$ is regular in E. Logarithmic differentiation gives

$$z\frac{F'(z)}{F(z)} = \alpha\left(1 + \frac{zf''(z)}{f'(z)}\right) + (1-\alpha)\frac{zf'(z)}{f(z)}. \tag{34}$$

Hence, if $f(z)$ is α-convex in E, then $F(z)$ is starlike in E.

Conversely, suppose that we start with an $F(z)$ in ST and try to solve equation (33) for $f(z)$, assuming that $\alpha \neq 0$. (The case $\alpha = 0$ gives $F(z) = f(z)$.) Formal manipulations with differential equations lead to the solution

$$f(z) = \left[\frac{1}{\alpha}\int_0^z \frac{[F(\zeta)]^{1/\alpha}}{\zeta}\, d\zeta\right]^{\alpha}. \tag{35}$$

Here again it is easy to show that if $\alpha > 0$, then it is possible to select branches so that $f(z)$ is regular in E, $f(0) = 0$, and $f'(0) = 1$. When $f(z)$ is defined by (35), then the right side of (33) gives $F(z)$. Further, the computation (34) shows that $f(z)$ is α-convex and hence univalent in E. We have

Theorem 6. Mocanu 1969 [2-B-990] and S. Miller 1973 [2-B-958]. If $\alpha > 0$, the two equations (33) and (35) give a one-to-one correspondence between the sets $\alpha - CV$ and ST, when the branches are selected properly.

In looking for the solutions of extremal problems in the set $\alpha - CV$, it is natural to examine the function $f(z)$ that corresponds to the Koebe function $F(z) = z/(1-z)^2$. Using this starlike function in (35) we obtain

$$k(z;\alpha) \equiv \left[\frac{1}{\alpha}\int_0^z \frac{\zeta^{1/\alpha}\,d\zeta}{\zeta(1-\zeta)^{2/\alpha}}\right]^\alpha, \tag{36}$$

a function that is called the *α-convex Koebe function.*

Numerical tables for this function have been computed (see Mocanu, Moldovan, and Reade 1974 [2-B-997]) and graphs of the boundary curves are shown for $\alpha = 1/8$, $1/2$, 1, 2, and $19/4$. In this paper the authors use $F(z) = z/(1+z)^2$ instead of $z/(1-z)^2$.

Theorem 7. PWO. S. Miller [2-B-958]. If $\alpha > 0$ and $f(z)$ is α-convex, then

$$k(-r;\alpha) \le |f(z)| \le k(r;\alpha), \qquad z = re^{i\theta}, \tag{37}$$

and both inequalities are sharp, for each $\alpha > 0$ and each r in $(0, 1)$.

The extreme terms in (37) can be expressed in terms of the hypergeometric functions.

Corollary. PLE. If $\alpha = 2$, we have

$$[\operatorname{Arc\,tan}\sqrt{r}]^2 \le |f(z)| \le \frac{1}{4}\left(\ln\frac{1+\sqrt{r}}{1-\sqrt{r}}\right)^2,$$

and if $\alpha > 2$, then $f(z)$ is a bounded function in E.

These results have been extended to functions with k-fold symmetry by Coonce and S. Miller 1974 [2-B-291].

By using variational methods Leach 1975 [2-B-808] proved a very general theorem concerning extreme values for functions in $\alpha - CV$, namely

Theorem 8. PWO. Suppose that $\alpha > 0$ and Φ is an entire function. Then for fixed z_0 in E, the maximum values for

$$\left|\Phi\left(\ln\frac{f(z_0)}{z_0}\right)\right| \quad \text{and} \quad \operatorname{Re}\Phi\left(\ln\frac{f(z_0)}{z_0}\right)$$

in the class $\alpha - CV$ are attained for some function of the form

$e^{-i\theta}k(e^{i\theta}z;\alpha)$ where $k(z;\alpha)$ is defined by equation (36). Here the case where $\Phi'(f(z_0)/z_0) = 0$ at an extreme point is excluded from consideration.

Of course, this theorem includes Theorem 7 as a special case. However, Theorem 8 does not cover the extreme values for $|a_n|$ when $f(z)$ has the form (3). Particular cases have been treated by several authors: S. Miller [2-B-958], J. Szynal 1972 [2-B-1417], and J. Szynal and Wajler 1974 [2-B-1418]. The bound for $|a_n|$ for all n has been investigated by Kulshrestha 1974 [2-B-778 and 2-B-779] and 1976 [2-B-780], and by Todorov [1]. The paper by Leach also contains some asymptotic results for $M(r, f)$, for each $\alpha \geq 0$.

Marx and Strohhäcker independently proved that if $f(z)$ is convex then $f(z)$ is starlike of order $1/2$ (see Problems 35 to 39). It is natural to ask for a bound on the order of starlikeness for an α-convex function. In this direction Eenigenburg and Nelson 1976 [3-B-230] have proved a rather surprising result, namely, for each α in (0, 1) the α-convex Koebe function (see equation (36)) is starlike of order zero *but is not starlike of order* ϵ *for any* $\epsilon > 0$. Hence $\beta(\alpha)$, the maximum order of starlikeness for α-convex functions, is not a continuous function of α but has a jump of $1/2$ at $\alpha = 1$.

For further work on this problem see S. Miller, Mocanu, and Reade 1978 [3-B-603], who prove that if $\alpha \geq 1$, then

$$\beta(\alpha) \equiv \inf_{z\in E} \operatorname{Re} \frac{zf'(z)}{f(z)} = \frac{\Gamma\left(\frac{1}{2} + \frac{1}{\alpha}\right)}{\sqrt{\pi}\,\Gamma\left(1 + \frac{1}{\alpha}\right)} \tag{38}$$

with equality for $f(z) = k(z;\alpha)$, the α-convex Koebe function.

Eenigenburg 1974 [2-B-353] has shown that if $\alpha \leq -2$ and $f(z)$ is in $\alpha - CV$, then either $f(z)$ is a bounded function or $f(z)$ is a rotation of the α-convex Koebe function with $\alpha = -2$.

Lewandowski, S. Miller, and Zlotkiewicz 1974 [3-B-531] have introduced the class of gamma-starlike functions, denoted here by γ-ST. A function is in γ-ST if it is regular in E, has the normalization (3), and satisfies

$$\operatorname{Re}\left[\left(\frac{zf'(z)}{f(z)}\right)^{1-\gamma}\left(1 + \frac{zf''(z)}{f'(z)}\right)^{\gamma}\right] \geq 0, \tag{39}$$

in E. Further, they assume that $f(z)$, $f'(z)$, and $1 + zf''(z)/f'(z)$ are not zero in $0 < |z| < 1$. For further work on functions that satisfy (39) see S. Miller 1976 [3-B-589].

Subsets of $\alpha - CV$ may be defined by demanding that the functional

$$J(\alpha, f(z)) \equiv \alpha Q_{CV}(z) + (1-\alpha)Q_{ST}(z)$$

lie in some special subset of the right half-plane $\text{Re } J > 0$. For some of the many possibilities see Chapter 13.3. Here we omit all references to papers that pursue this line of research.

3. ALPHA-SPIRAL FUNCTIONS

Suppose that we modify the condition $\text{Re } (zf'(z)/f(z)) > 0$ by inserting a factor $e^{i\alpha}$ to obtain

$$\text{Re } e^{i\alpha} \frac{zf'(z)}{f(z)} > 0. \tag{40}$$

If $f(z)$ has the usual normalization, then near $z = 0$ we have

$$e^{i\alpha} \frac{zf'(z)}{f(z)} = \cos \alpha + i \sin \alpha + \sum_{n=1}^{\infty} c_n z^n. \tag{41}$$

It follows from (40) that $\cos \alpha \geq 0$ (put $z = 0$) and hence $|\alpha| \leq \pi/2$. If $\alpha = \pm\pi/2$, then $f(z) \equiv z$, so henceforth we always assume that $|\alpha| < \pi/2$. If we normalize the expression on the right side of (41), we arrive at

$$\frac{1}{\cos \alpha}\left[e^{i\alpha} \frac{zf'(z)}{f(z)} - i \sin \alpha\right] = 1 + \sum_{n=1}^{\infty} p_n z^n \equiv p(z) \tag{42}$$

where $p(z)$ is a function in P. Conversely, if $p(z)$ is in P, then the inequality (40) is satisfied in E.

Now suppose that we set $p(z) = (1+z)/(1-z)$ in (42) and solve for $f(z)$. Using $F(z)$ for the solution in this particular case, we find that

$$F(z) = \frac{z}{(1-z)^{2s}}, \qquad \text{where} \qquad s = e^{-i\alpha} \cos \alpha, \tag{43}$$

the *α-spiral Koebe function.*

Definiton 4. If $f(z)$ has the normaliztion (3), is regular in E, and satisfies the inequality (40) in E, then it is said to be *α-spiral* in E (or merely α-spiral). The set of all such funtions will be denoted by $SP(\alpha)$. The set $SP \equiv \bigcup SP(\alpha)$ over all α in $[-\pi/2, \pi/2]$.

Montel 1933 [1-B-939, p. 15] seems to be the first to suggest the name spiral (or spirallike) for functions in SP. A reason for this name will be given in Problems 60 through 68.

The same computation that gave (43) will also give a representation formula for the class $SP(\alpha)$.

Theorem 9. PLE. Špaček 1932 [1-B-1317]. The function $f(z)$ is in $SP(\alpha)$ if and only if there is a $p(z)$ in P such that

$$f(z) = ze^{I(z)} \tag{44}$$

where

$$I(z) = (e^{-i\alpha} \cos \alpha) \int_0^z \frac{p(\zeta) - 1}{\zeta} d\zeta. \tag{45}$$

Of course, we need

Theorem 10. Špaček [1-B-1317]. If $f(z)$ is in $SP(\alpha)$, then $f(z)$ is univalent in E.

Proof. If $f(z)$ is not univalent in E, then there is some circle $|z| = r < 1$ on which $f(z)$ takes the same value at two distinct points.

Assume that $f(re^{it_1}) = f(re^{it_2})$ where $t_1 \not\equiv t_2 \pmod{2\pi}$ and r is fixed in $(0, 1)$. Now $f(z)$ has only the one zero at $z = 0$ in E, hence $\Delta \arg f(C_r) = 2\pi$. Therefore, t_1 and t_2 can be selected so that $0 \leq t_1 < t_2 < 2\pi$ and Γ, the image of $z = re^{i\theta}$ for $t_1 \leq \theta \leq t_2$, is a simple closed curve that does *not* enclose the origin. Hence $\ln w = \ln f(z)$ is single-valued on Γ and $\ln f(re^{it_1}) = \ln f(re^{it_2})$. Consequently,

$$0 = \int_\Gamma \frac{d}{dz} \ln f(z)\, dz = \int_{t_1}^{t_2} \frac{f'(z)}{f(z)} \frac{dz}{d\theta}\, d\theta$$

$$= i \int_{t_1}^{t_2} z \frac{f'(z)}{f(z)}\, d\theta, \qquad z = re^{i\theta}.$$

From the inequality (40), the values of the integrand lie in a half-plane bounded by a line through the origin. Therefore, just as in the proof of the Noshiro-Warschawski Theorem (Theorem 7.13), the integral cannot vanish. ■

We can use the technique of dominant power series (see Chapter 7.3) to obtain bounds for the coefficients. Indeed,

$$p(z) << \frac{1+z}{1-z}$$

and so

$$\int_0^z \frac{p(t)-1}{t}\,dt << \int_0^z \frac{1}{t}\left(\frac{1+t}{1-t}-1\right)dt = -2\ln(1-z).$$

Therefore, from (44) and (45)

$$f(z) << ze^{(\cos\alpha)(-2\ln(1-z))} = \frac{z}{(1-z)^{2\cos\alpha}}. \tag{46}$$

Using the binomial theorem we find that if

$$f(z) = z + \sum_{n=2}^{\infty} a_n z^n \tag{3}$$

is in $SP(\alpha)$, then

$$|a_n| \le \frac{2\cos\alpha(1+2\cos\alpha)\cdots(n-2+2\cos\alpha)}{(n-1)!} \le n\cos\alpha, \tag{47}$$

with equality on the extreme right only when $n = 2$, or $\alpha = 0$.

Unfortunately, this bound is *not* sharp and the method fails for two reasons. First, the function on the right side of (46) is not in $SP(\alpha)$ if $\alpha \neq 0$, and, second, the method ignores the effect of the complex nature of the multiplier $e^{-i\alpha}\cos\alpha$ in (45).

These difficulties were conquered by Zamorski 1960 [1-B-1507], who used a method due to Clunie 1959 [1-B-216] (see Chapter 17.3). It should be observed that Zamorski uses the multiplier $1/(1-ai)$ in place of $s = e^{-i\alpha}\cos\alpha$. Consequently, we have $a = -\tan\alpha$ in Zamorski's work.

Theorem 11. If $f(z)$ is in $SP(\alpha)$, then

(48)
$$|a_n| \le \prod_{k=1}^{n-1} \frac{|k+2s-1|}{k} = \frac{1}{(n-1)!} \prod_{k=1}^{n-1} ((k-1)^2 + 4k\cos^2\alpha)^{1/2},$$

and this inequality is sharp for each $n \ge 2$. An extremal function is given by (43) or any rotation of $F(z)$.

Proof. From (44) and (45) (or from (42)), we have

(49)
$$z\frac{f'(z)}{f(z)} = 1 + s(p(z) - 1).$$

Since $p(z)$ is in P, there is a function $b(z)$ that satisfies the conditions of Schwarz's Lemma and such that

(50)
$$p(z) = \frac{1+b(z)}{1-b(z)} \qquad \text{or} \qquad p(z) - 1 = \frac{2b(z)}{1-b(z)}.$$

Using this in (49), we can obtain

$$b(z)[zf'(z) + (2s-1)f(z)] = zf'(z) - f(z).$$

If $b(z) = \Sigma_{k=1}^{\infty} b_k z^k$ and $f(z)$ is given by (3), we have

(51)
$$\left(\sum_{k=1}^{\infty} b_k z^k\right)\left(\sum_{k=1}^{\infty} (k+2s-1)a_k z^k\right) = \sum_{k=1}^{\infty} (k-1)a_k z^k.$$

One of the tricks employed by Clunie is to break the second sum on the left side into two parts, $\Sigma_{k=1}^{n-1}$ and $\Sigma_{k=n}^{\infty}$, and to observe that on multiplication by $b(z)$ the second sum gives terms with exponents that exceed n. We transpose such terms to the right side and, by the definition of c_k, for $k \ge n+1$, we have

(52)
$$\left(\sum_{k=1}^{\infty} b_k z^k\right)\left(\sum_{k=1}^{n-1} (k+2s-1)a_k z^k\right) = \sum_{k=1}^{n} (k-1)a_k z^k + \sum_{k=n+1}^{\infty} c_k z^k,$$

where, fortunately, it turns out that we do not need exact expressions for the c_k. We now set $z = re^{i\theta}$ and integrate on θ. Eventually we will let

$r \to 1^-$, and to save excessive writing, we set $r = 1$ at the outset. Then (see Libera [2-B-852])

$$\begin{aligned}
\sum_{k=1}^{n-1} |k+2s-1|^2 |a_k|^2 &= \frac{1}{2\pi}\int_0^{2\pi} \left| \sum_{k=1}^{n-1} (k+2s-1)a_k z^k \right|^2 d\theta \\
&\geq \frac{1}{2\pi}\int_0^{2\pi} \left| \sum_{k=1}^{n-1} (k+2s-1)a_k z^k \right|^2 |b(z)|^2 \, d\theta \\
&\geq \frac{1}{2\pi}\int_0^{2\pi} \left| b(z) \sum_{k=1}^{n-1} (k+2s-1)a_k z^k \right|^2 d\theta \\
&\geq \frac{1}{2\pi}\int_0^{2\pi} \left| \sum_{k=1}^{n} (k-1)a_k z^k + \sum_{k=n+1}^{\infty} c_k z^k \right|^2 d\theta \\
&\geq \sum_{k=1}^{n} (k-1)^2 |a_k|^2 + \sum_{k=n+1}^{\infty} |c_k|^2 \\
&\geq \sum_{k=1}^{n} (k-1)^2 |a_k|^2 .
\end{aligned}$$

The extremes in this sequence of inequalities give

$$\sum_{k=1}^{n} (k-1)^2 |a_k|^2 \leq \sum_{k=1}^{n-1} |k+2s-1|^2 |a_k|^2$$

or

$$(n-1)^2 |a_n|^2 \leq \sum_{k=1}^{n-1} [|k+2s-1|^2 - (k-1)^2] |a_k|^2$$

or

$$|a_n|^2 \leq \frac{4\cos^2\alpha}{(n-1)^2} \sum_{k=1}^{n-1} k|a_k|^2. \tag{53}$$

Since $a_1 = 1$, equation (53) gives the bounds

$$|a_2| \leq 2\cos\alpha \qquad \text{and} \qquad |a_3| \leq \cos\alpha(1+8\cos^2\alpha)^{1/2}, \tag{54}$$

and these agree with the bounds given in the theorem. We now use mathematical induction to prove that the bound (48) is correct for all $n \geq 2$.

We assume that for $m = 2, 3, \ldots, n-1$

$$|a_m| \leq \prod_{k=1}^{m-1} \frac{|k+2s-1|}{k}$$

and that when $|a_k|$ is replaced by this bound in (53), we get equality. Using this when $m = n-1$, we have

$$\begin{aligned} |a_{n-1}|^2 &\leq \prod_{k=1}^{n-2} \left(\frac{|k+2s-1|}{k}\right)^2 \\ &= \frac{4\cos^2\alpha}{(n-2)^2}\left[1 + \sum_{k=2}^{n-2} k \prod_{j=1}^{k-1} \left(\frac{|j+2s-1|}{j}\right)^2\right]. \end{aligned} \tag{55}$$

For a_n, equation (53) and the induction hypothesis give

$$\begin{aligned} |a_n|^2 \leq {} & \frac{4\cos^2\alpha}{(n-1)^2}\left[1 + \sum_{k=2}^{n-2} k \prod_{j=1}^{k-1} \left(\frac{|j+2s-1|}{j}\right)^2\right] \\ & + \frac{4\cos^2\alpha}{(n-1)^2}(n-1) \prod_{k=1}^{n-2} \left(\frac{|k+2s-1|}{k}\right)^2. \end{aligned}$$

When equation (55) is applied to the sum, we find that

$$\begin{aligned} |a_n|^2 \leq {} & \frac{(n-2)^2}{(n-1)^2} \prod_{k=1}^{n-2} \left(\frac{|k+2s-1|}{k}\right)^2 \\ & + \frac{4\cos^2\alpha}{(n-1)^2}(n-1) \prod_{k=1}^{n-2} \left(\frac{|k+2s-1|}{k}\right)^2 \end{aligned}$$

$$|a_n|^2 \leq \prod_{k=1}^{n-2} \left(\frac{|k+2s-1|}{k}\right)^2 \left[\frac{(n-2)^2 + 4(n-1)\cos^2\alpha}{(n-1)^2}\right].$$

Since the term in brackets is $[(k-1)^2 + 4k\cos^2\alpha]/k^2$ when $k = n-1$, we obtain (48).

To see that the bounds are sharp, we observe that the function $F(z) = z/(1-z)^{2s}$ is in $SP(\alpha)$ for each α with $|\alpha| < \pi/2$, and

$$\frac{z}{(1-z)^{2s}} = z + \sum_{n=2}^{\infty} \prod_{k=1}^{n-1} \frac{k+2s-1}{k} z^n. \tag{56}$$ ■

This work can be extended by refining the class $SP(\alpha)$. Indeed, suppose that $0 < \varrho < 1$. We let $SP(\alpha, \varrho)$ be the subset of $SP(\alpha)$ of functions for which

(57) $$\operatorname{Re} e^{i\alpha} Q_{ST}(f) > \varrho \cos\alpha, \qquad |\alpha| < \pi/2,$$

for all z in E. A normalized function that satisfies (57) is called *α-spiral of order ϱ.*

Libera 1967 [2-B-852] proved that for this class we have the sharp coefficient bound

$$|a_n| \leq \prod_{k=1}^{n-1} \frac{|k + 2s(1-\varrho) - 1|}{k}, \qquad s = e^{-i\alpha}\cos\alpha.$$

An extremal function is $F(z) = z/(1-z)^{2s(1-\rho)}$.

Pinchuk 1968 [2-B-1122] obtained sharp upper and lower bounds for $|f(z)|$, $\ln|f(z)/z|$, and $\arg f(z)/z$ for $f(z)$ in $SP(\alpha, \varrho)$ Note that in (57) Pinchuk uses ϱ in place of $\varrho\cos\alpha$. See R. S. L. Srivastava and S. K. Bajpai 1972 [3-B-881] and Kulshrestha 1973 [2-B-775].

Other subsets of $SP(\alpha)$ have been studied by Mogra and Ahuja [1].

Whenever a new class of univalent functions is introduced, one naturally tries to find an Alexander type theorem. This means that if $F(z)$ is in one class, we look for a normalized function $f(z)$ such that $F(z) = zf'(z)$ or

$$f(z) = \int_0^z \frac{F(\zeta)}{\zeta}\, d\zeta.$$

Following this idea Robertson 1969 [2-B-1221] defines a new class $CVSP(\alpha)$ of *convex α-spiral functions* by the condition $F(z)$ is in $SP(\alpha)$ if and only if $f(z)$ is in $CVSP(\alpha)$. Thus $CVSP(\alpha)$ consists of those normalized functions, regular in E, for which

(58) $$\operatorname{Re}\left(e^{i\alpha} Q_{CV}(f)\right) > 0,$$

for all z in E. As usual $CVSP = \cup_\alpha CVSP(\alpha)$. Then an Alexander type theorem becomes true for the classes $SP(\alpha)$ and $CVSP(\alpha)$ by definition. However, the reader must be aware that if $f(z)$ is in $CVSP(\alpha)$, it does not follow that $f(E)$ is convex or even spirallike in shape. Even more surprising is the fact that $f(z)$ need not be univalent as the example $f(z) = i(1-z)^i - i = z + \cdots$ shows (see Problems 47 and 48).

Robertson proved that if α is suitably restricted and $f(z)$ is in $CVSP(\alpha)$, then $f(z)$ is univalent in E, and he obtained the interval $0 \leq \cos\alpha < 0.2315$ as one that implies univalence. Using a different method Libera and Ziegler 1972 [2-B-858] extended this interval to

$0 \le \cos\alpha < 0.2564$. Chichra 1975 [2-B-268] extended it still further to $0 \le \cos\alpha < (\sqrt{6}-\sqrt{2})/4 \approx 0.2588$. Finally, Pfaltzgraff 1975 [2-B-1111] proved that if $0 \le \cos\alpha \le 1/2$ and $f(z)$ is in $CVSP(\alpha)$, then $f(z)$ is univalent in E. This is best possible because, as Robertson 1969 [2-B-1221] proved, for each α, with $1/2 < \cos\alpha < 1$, there is an $f(z)$ in $CVSP(\alpha)$ that is not univalent in E.

Pinchuk 1971 [2-B-1124, p. 11] introduced and studied the set $CVSP(\alpha, \beta)$ of normalized functions regular in E for which $\operatorname{Re} e^{i\alpha} Q_{CV}(z) > \beta$ for all z in E. Here we assume that $|\alpha| \le \pi/2$. Further, $\cos\alpha \ge \beta$, for otherwise the set $CVSP(\alpha, \beta)$ is empty.

The inequality (57) can be modified by imposing the condition

$$\left| \frac{e^{i\alpha} Q_{ST}(f) - i \sin\alpha}{\cos\alpha} - M \right| < M, \qquad M > 1/2, \tag{59}$$

where as usual $f(z)$ is normalized and regular in E.

Kulshrestha 1976 [3-B-488] calls functions in $CVSP(\alpha)$ *Robertson functions.* Those that satisfy (59) with $Q_{ST}(f)$ replaced by $Q_{CV}(f)$ are called *bounded Robertson functions.*

Ahuja [1] introduced and studied the class of λ-Robertson functions of order α and type β. These are the normalized functions regular in E, for which

$$\left| \frac{Q_{CV}(f) - 1}{(2\beta - 1) Q_{CV}(f) + 2\beta(1-\alpha) e^{-i\lambda} \cos\lambda + 1 - 2\beta} \right| < 1$$

for all z in E. Here $-\pi/2 < \lambda < \pi/2$, $0 \le \alpha < 1$ and $0 < \beta \le 1$. Special values of the parameters give subclasses that were studied earlier. Hence, Ahuja's results contain many of the earlier results as special cases.

4. OTHER CENTERS FOR A STARLIKE FUNCTION

A convex domain is convex for every choice of the origin. However, a domain that is starlike with respect to one point w_0 may fail to be starlike with respect to a second point w_1. Of course, if a domain D is bounded and w_0 lies outside D, then it is impossible for $\arg(w - w_0)$ to be nondecreasing as w traverses the boundary of D in a counterclockwise direction.

Suppoe that D is a bounded domain and the boundary can be parametrized by $w(\theta)$ where $w(\theta)$ is a continuous function of θ with period 2π. Then D is said to be starlike with respect to a boundary point $w_0 = w(\theta_0)$ if $\arg(w(\theta) - w_0)$ is a nondecreasing function of θ for $\theta_0^+ \leq \theta \leq (\theta_0 + 2\pi)^-$. We will return to this situation in Section 6.

Suppose next that $f(z)$ is a bounded univalent function and w_0 lies inside $D = f(E)$. Then $w_0 = f(\zeta)$ for some unique $\zeta = \varrho e^{i\theta_0}$. If $\varrho < r < 1$,then the domain $D_r = f(E_r)$ is starlike with respect to $w_0 = f(\zeta)$ if and only if

$$\text{(60)} \qquad \operatorname{Re} \frac{zf'(z)}{f(z) - f(\zeta)} \geq 0, \qquad z \in C_r\colon |z| = r,$$

as we proved in Chapter 8 (see equation (8.7)).

We may expect that if the inequality (60) is true for $r = 1$, then it is true for all r in $\varrho < r < 1$, and one might try to base a proof on Ford's Theorem (Theorem 8.3). However, a difficulty intrudes because of the pole at $z = \zeta$. This is not an accident. To the contrary it turns out that the inequality (60) may be true for $r = 1$ and *false* for $1 - \epsilon < r < 1$ for some positive ϵ. An example of this type of behavior will be given in Chapter 10.4 in connection with functions convex in the direction of the imaginary axis.

If the domain $D = f(E)$ is unbounded, then we say that D is starlike with respect to w_0 if every ray from w_0 either meets D in a segment or is contained in D.

In this definition, w_0 may be outside D, but if w_0 is in the closure of D, then it is easy to see that the inequality (60) will hold at all points $z = e^{i\theta}$ where the quantities involved are well-defined.

Egervâry 1937 [1-B-274] studied the mapping properties of the Cesàro means of the partial sums of the geometric series $z/(1-z) = z + z^2 + \cdots + z^n + \cdots$. The Cesàro means of the first, second, and third order are (by definition)

$$\text{(61)} \quad S_n^{(1)}(z) \equiv \frac{1}{n}[nz + (n-1)z^2 + (n-2)z^3 + \cdots + 2z^{n-1} + z^n]$$

$$= \frac{1}{n} \sum_{k=1}^{n} \binom{n+1-k}{1} z^k,$$

$$\text{(62)} \quad S_n^{(2)}(z) \equiv \frac{2}{n(n+1)} \left[\binom{n+1}{2} z + \binom{n}{2} z^2 + \cdots + \binom{3}{2} z^{n-1} + \binom{2}{2} z^n \right]$$

$$= \frac{2}{n(n+1)} \sum_{k=1}^{n} \binom{n+2-k}{2} z^k,$$

and

$$(63) \quad S_n^{(3)}(z) \equiv \frac{6}{n(n+1)(n+2)}\left[\binom{n+2}{3}z + \binom{n+1}{3}z^2 + \cdots + \binom{3}{3}z^n\right]$$

$$= \frac{6}{n(n+1)(n+2)} \sum_{k=1}^{n} \binom{n+3-k}{3} z^k.$$

Egerváry proved the following three theorems about the Cesàro means of the geometric series. See also Fejér and Szegö 1951 [1-B-301].

Theorem 12. PWO. If $n \geq 2$, the Cesàro mean of first order, $S_n^{(1)}(z)$, is univalent in E and the image is starlike with respect to the boundary point $S_n^{(1)}(1)$.

Theorem 13. PWO. If $n \geq 2$, the Cesàro mean of second order, $S_n^{(2)}(z)$, is univalent in E and is starlike with respect to the origin (and perhaps other points as well).

Theorem 14. PWO. If $n \geq 2$, the Cesàro mean of third order, $S_n^{(3)}(z)$, is univalent and convex.

As Egerváry remarks, these three theorems indicate the smoothing effect of Cesàro means as k increases.

We must emphasize that theorems 12, 13, and 14 apply only to the Cesàro means of the geometric series. It would be of interest to see if similar theorems hold for the Cesàro means of an arbitrary univalent function.

Robertson 1936 [1-B-1145] proved that if every partial sum

$$s_n(z) = z + a_2z^2 + \cdots + a_nz^n, \qquad n = 1, 2, 3, \ldots,$$

is univalent in E, then the same is true for each Cesàro sum of the first order. Bustoz 1975 [2-B-200] proves the same result and relates it to certain sums with Jacobi polynomials $P_k^{(\beta,\beta)}(x)$ as coefficients.

In the next seven theorems we investigate functions for which the domain $D = f(E)$ is starlike with respect to $w_0 = f(\zeta)$ where ζ is a fixed point with $0 < |\zeta| = \varrho < 1$. Following Wald 1978 [3-B-962] we will use a trick to remove the pole at $z = \zeta$ in

(64) $$p(z) \equiv \frac{zf'(z)}{f(z) - f(\zeta)}$$

by transferring ζ to the origin. First, some notation: We let $ST(\zeta)$ be the set of all functions $f(z)$ regular in E, with the usual normaliztion, and such that $f(E)$ is starlike with respect to $f(\zeta)$. This notation may seem to be confusing because $ST(\alpha)$ denotes the set of functions starlike of order α, but the difference in the two sets lies in the use of the different letters ζ and α.

It will be useful to adopt a slightly different normalization and ask that $f'(\zeta) = 1$. Thus, we let $STN(\zeta)$ be the set of functions

(65) $$f(z) = \sum_{n=1}^{\infty} b_n z^n$$

which are regular in E and for which $f'(\zeta) = 1$, and $f(E)$ is starlike with respect to $f(\zeta)$.

The set $STN(\zeta)$ will lead naturally to a related set, $P(\zeta)$. Here $P(\zeta)$ is the set of all functions

(66) $$q(z) = \sum_{n=0}^{\infty} A_n z^n \equiv 1 + \sum_{n=1}^{\infty} B_n (z - \zeta)^n$$

that are regular in E, have positive real part in E, and for which $q(\zeta) = 1$.

The following seven theorems are due to Wald [3-B-962].

Theorem 15. If $q(z)$ is in $P(\zeta)$ and has the series expansions (66), then for all $n \geq 1$,

(67) $$|A_0| \leq \frac{1+\varrho}{1-\varrho}, \qquad |A_n| \leq 2\,\frac{1+\varrho}{1-\varrho}, \qquad \varrho = |\zeta|,$$

and

(68) $$|B_n| \leq \frac{2}{(1+\varrho)(1-\varrho)^n}.$$

All the inequalities are sharp.

Proof. We will use the two mappings of E onto itself,

(69) $$L(z) \equiv \frac{z+\zeta}{1+\bar{\zeta}z} \qquad \text{and} \qquad M(z) \equiv \frac{z-\zeta}{1-\bar{\zeta}z}.$$

Each one is the inverse of the other, so $L(M(z)) = M(L(z)) = z$.

There is a one-to-one correspondence between the sets P and $P(\zeta)$. If p and q are related by

$$q(L(z)) = p(z) \qquad \text{or} \qquad q(z) = p(M(z)),$$

then (as the reader can easily check) $p(z)$ is in P if and only if $q(z)$ is in $P(\zeta)$.

We recall the Herglotz representation for a function in P, and to simplify writing we set $x = e^{-i\phi}$. Then from Theorem 7.22

$$\text{(70)} \qquad p(z) = \frac{1}{2\pi}\int_0^{2\pi} \frac{1+xz}{1-xz}\, d\mu(\phi) \equiv \frac{1}{2\pi}\int_0^{2\pi} G(z, x)\, d\mu(\phi),$$

where $\mu(\phi)$ is in $M(0, 2\pi)$. Consequently, any function $q(z)$ in $P(\zeta)$ can be obtained as a Stieltjes integral by using $G(M(z), x)$ as the integrand in (70). Now a brief computation shows that

$$\text{(71)} \qquad G(M(z), x) = \frac{1-\bar{\zeta}z+xz-x\zeta}{1-\bar{\zeta}z-xz+x\zeta} = \frac{1-x\zeta+z(x-\bar{\zeta})}{1+x\zeta-z(x+\bar{\zeta})}$$

$$= \frac{1-x\zeta}{1+x\zeta}\cdot\frac{1+\dfrac{z(x-\bar{\zeta})}{1-x\zeta}}{1-\dfrac{z(x+\bar{\zeta})}{1+x\zeta}} \equiv \sum_{n=0}^{\infty} a_n z^n$$

where

$$a_0 = \frac{1-x\zeta}{1+x\zeta}$$

and for $n \geq 1$

$$a_n = \frac{1-x\zeta}{1+x\zeta}\left[\left(\frac{x+\bar{\zeta}}{1+x\zeta}\right)^n + \left(\frac{x-\bar{\zeta}}{1-x\zeta}\right)\left(\frac{x+\bar{\zeta}}{1+x\zeta}\right)^{n-1}\right].$$

But for $|x| = 1$ and $|\zeta| = \varrho < 1$, each term in paraentheses represents a mapping of the unit circle onto itself. Hence $|a_n| \leq 2(1+\varrho)/(1-\varrho)$.

Then on integration, as dictated by equation (70), we get (67). Equality can occur if $x = 1$ and $\zeta = -\varrho$. In that case, a proper choice of $\mu(\phi)$ gives

$$q(z) = \frac{1+\varrho}{1-\varrho}\cdot\frac{1+z}{1-z}, \qquad q(-\varrho) = 1.$$

For (68) we return to (71) and note that $G(M(z), x)$ can be put in the form

$$G(M(z), x) = \frac{1 - \bar{\zeta}(z-\zeta) - \varrho^2 + x(z-\zeta)}{1 - \bar{\zeta}(z-\zeta) - \varrho^2 - x(z-\zeta)}$$

$$= \frac{1 - \varrho^2 + (z-\zeta)(x-\bar{\zeta})}{1 - \varrho^2 - (z-\zeta)(x+\bar{\zeta})} \equiv \sum_{n=0}^{\infty} b_n (z-\zeta)^n,$$

where $b_0 = 1$, and for $n \geq 1$

$$b_n = \frac{1}{(1-\varrho^2)^n}[(x+\bar{\zeta})^n + (x-\bar{\zeta})(x+\bar{\zeta})^{n-1}] = \frac{2x}{(1-\varrho^2)^n}(x+\bar{\zeta})^{n-1}.$$

We observe that the infinite series converges for $|z-\zeta| < 1-\varrho$. Since $|b_n| \leq 2(1+\varrho)^{n-1}/(1-\varrho^2)^n = 2/(1+\varrho)(1-\varrho)^n$, the integral in (70) gives (68).

Equality can occur if $x = 1$ and $\zeta = \varrho$. Then, just as in the first part of the proof, we find that

$$q(z) = \frac{1-\varrho}{1+\varrho} \cdot \frac{1+z}{1-z}, \qquad q(\varrho) = 1. \qquad \blacksquare$$

Theorem 16. PLE. Let $f(z) = z + \Sigma_{n=2}^{\infty} a_n z^n$ be regular in E. Then $f(z)$ is in $ST(\zeta)$ if and only if

$$g(z) \equiv \frac{f(L(z)) - f(\zeta)}{(1-\varrho^2)f'(\zeta)} \tag{72}$$

is in ST.

Theorem 17. Let $g(z) = z + \Sigma_{n=2}^{\infty} d_n z^n$ be regular in E. Then $g(z)$ is in ST if and only if

$$f(z) \equiv \frac{g(M(z)) - g(-\zeta)}{(1-\varrho^2)g'(-\zeta)} \tag{73}$$

is in $ST(\zeta)$.

Proof. One can deduce this directly from equation (73). Perhaps it is instructive to obtain Theorem 17 from Theorem 16. From equation (72) we have

$$g(-\zeta) = \frac{-f(\zeta)}{(1-\varrho^2)f'(\zeta)}$$

and

$$g'(-\zeta) = \frac{1}{(1-\varrho^2)^2 f'(\zeta)}\,.$$

If in (72) we replace z by $M(z)$ and use the above expressions for $f(\zeta)$ and $f'(\zeta)$, we will obtain (73). ■

We now look at the set $STN(\zeta)$ normalized by the condition $F'(\zeta) = 1$. Theorems 16 and 17 give immediately

Theorem 18. PLE. If $F(z)$ is in $STN(\zeta)$, then

$$g(z) \equiv \frac{F(L(z)) - F(\zeta)}{1-\varrho^2}$$

is in ST, Further, if $g(z)$ is in ST, then

$$F(z) \equiv (1-\varrho^2)(g(M(z)) - g(-\zeta))$$

is in $STN(\zeta)$.

Theorem 19. Suppose that

(3) $$f(z) = z + \sum_{n=2}^{\infty} a_n z^n$$

is regular in E. Then $f(z)$ is in $ST(\zeta)$ if and only if in E

(74) $$\operatorname{Re} Q_1(z) \equiv \operatorname{Re} \frac{(z-\zeta)(1-\bar{\zeta}z)f'(z)}{(1-\varrho^2)(f(z)-f(\zeta))} > 0, \qquad Q_1(\zeta) = 1.$$

Observe that in the inequality (74) the singularity at $z = \zeta$ is a removable one.

Proof. By Theorem 16

(72) $$g(z) = \frac{f(L(z)) - f(\zeta)}{(1-\varrho^2)f'(\zeta)}$$

is in ST if and only if $f(z)$ is in $ST(\zeta)$. Further, $g(z)$ is in ST if and only if $\operatorname{Re} zg'(z)/g(z) > 0$ in E. Now

$$\operatorname{Re}\frac{zg'(z)}{g(z)} = \operatorname{Re}\left(\frac{zf'(L(z))}{f(L(z)) - f(\zeta)} \cdot \frac{1-\varrho^2}{(1+\bar{\zeta}z)^2}\right). \tag{75}$$

Since $M(z)$ maps E onto E, we can replace z by $M(z)$ on the right side of (75). A brief computation shows that we have

$$\operatorname{Re}\left(\frac{M(z)f'(z)}{f(z)-f(\zeta)} \cdot \frac{1-\varrho^2}{(1+\bar{\zeta}M(z))^2}\right) = \operatorname{Re}\frac{(z-\zeta)(1-\bar{\zeta}z)f'(z)}{(1-\varrho^2)(f(z)-f(\zeta))}. \qquad \blacksquare$$

If c is a suitable constant so that $cf(z) = F(z)$ is a member of $STN(\zeta)$, then we can multiply the top and bottom of the expression in (74) to obtain

Theorem 20. If $F(z) = \sum_{n=1}^{\infty} b_n z^n$ and $F'(\zeta) = 1$, then $F(z)$ is in $STN(\zeta)$ if and only if in E

$$\operatorname{Re} Q_2(z) \equiv \operatorname{Re}\frac{(z-\zeta)(1-\bar{\zeta}z)F'(z)}{(1-\varrho^2)(F(z)-F(\zeta))} > 0, \qquad Q_2(\zeta) = 1. \tag{76}$$

Now that Theorems 19 and 20 have been proved by a somewhat long argument, and we have found the proper form for $Q_1(z)$ and $Q_2(z)$ in equations (75) and (76), a shorter and quicker approach presents itself.

On $|z| = 1$, it is clear that

$$S(z) \equiv \frac{(z-\zeta)(1-\bar{\zeta}z)}{z(1-\varrho^2)} \tag{77}$$

is a real number. The starlike condition suggests that

$$\operatorname{Re} p(z) \equiv \operatorname{Re}\frac{zf'(z)}{f(z)-f(\zeta)} \geq 0$$

for $\varrho < |z| < 1$. The condition (74) can be put in the form

$$\operatorname{Re}(S(z)p(z)) > 0. \tag{78}$$

Now assume that $f'(z)$ is continuous on $|z| = 1$, and $f(E)$ is starlike with respect to ζ. Then the left side of (78) is nonnegative for $|z| = 1$. But the left side of (78) is harmonic in E and hence (78) is true in E. This gives Theorem 19 (and also 20) when $f'(z)$ is continuous for

$|z| = 1$. It seems like a simple matter to use the same argument on a smaller disk $z \in E_r$, $\varrho < r < 1$, and take the limit as $r \to 1$, but such an approach has a slight flaw: it assumes that if $f(E)$ is starlike with respect to ζ, then $f(E_r)$ is also, and this has not been proved. We will see in Chapter 10.4 that such an assertion is false.

It is natural to ask about the coefficients in equation (3) when $f(z)$ is in $ST(\zeta)$. If $0 < \zeta < 1$, then we can do no better than $|a_n| \le n$, because $k(z) \equiv z/(1-z)^2$ is starlike with respect to $f(\zeta)$ for each ζ in $(0, 1)$. If ζ is not a real number, then we have the same sharp bound for any ζ in E, by a rotation of the Koebe function. Using a different method Ozaki 1941 [1-B-1022, p. 79] and De Bruijn 1941 [1-B-235] have proved that if $f(E)$ is starlike with respect to any w_0 and $f(z)$ is in S, then $|a_n| \le n$ for each $n \ge 2$. Their method will be covered in Problems 10.58 through 10.61.

Theorem 21. Suppose that $F(z)$ is in $STN(\zeta)$ and has the Taylor series

$$F(z) = F(\zeta) + \sum_{n=1}^{\infty} c_n(z-\zeta)^n. \tag{79}$$

Then for each $n \ge 1$

$$|c_n| \le \frac{n+\varrho}{(1+\varrho)(1-\varrho)^{n-1}} \equiv C_n, \tag{80}$$

and this result is sharp for each n.

Proof. From inequality (76) and Theorem 15 we have

$$\frac{(z-\zeta)(1-\bar{\zeta}z)F'(z)}{(1-\varrho^2)(F(z)-F(\zeta))} \ll 1 + \sum_{n=1}^{\infty} \frac{2}{(1+\varrho)(1-\varrho)^n}(z-\zeta)^n. \tag{81}$$

For brevity set $Z = z - \zeta$. Then (for sufficiently small Z) (81) gives

$$\frac{F'(z)}{F(z)-F(\zeta)} \ll \frac{1-\varrho^2}{Z(1-\varrho^2-\varrho Z)}\left[1 + \frac{2}{1+\varrho}\cdot\frac{Z}{1-\varrho-Z}\right].$$

Using partial fractions, we find that

$$\frac{F'(z)}{F(z)-F(\zeta)} \ll \frac{1}{Z} - \frac{\varrho}{1-\varrho^2-\varrho Z} + \frac{2}{1-\varrho-Z}. \tag{82}$$

Direct integration of (82) will yield an erroneous result because of the singularity at $Z = 0$. However, if we integrate both sides from Z_0 to Z and then let $Z_0 \to 0$, we obtain a correct result, namely

$$\ln \frac{F(z) - F(\zeta)}{z - \zeta} << \ln \frac{1 - \varrho^2 - \varrho Z}{1 - \varrho^2} - 2 \ln \frac{1 - \varrho - Z}{1 - \varrho}.$$

Hence

$$\begin{aligned} F(z) - F(\zeta) &<< Z \frac{1 - \varrho Z/(1 - \varrho^2)}{[1 - Z/(1 - \varrho)]^2} \\ &<< (z - \zeta) + \sum_{n=2}^{\infty} \frac{n + \varrho}{(1 + \varrho)(1 - \varrho)^{n-1}} (z - \zeta)^n. \end{aligned}$$

This gives the bounds (80) stated in the theorem.

To see that these bounds are sharp, let $\zeta = \varrho > 0$ and set

$$\Phi(z) = \Phi(\zeta) + (z - \zeta) \frac{1 - \dfrac{\varrho}{1 - \varrho^2}(z - \zeta)}{\left(1 - \dfrac{z - \zeta}{1 - \varrho}\right)^2}.$$

Then $\Phi(z)$ has the coefficients C_n defined in (80). Further, $\Phi(z)$ satisfies (82) with = in place of $<<$. A little effort will show that the left side of (81) with F replaced by Φ gives $(1 - \varrho)(1 + z)/(1 + \varrho)(1 - z)$, which has positive real part in E. ■

I. A. Aleksandrov 1959 [1-B-30] has considered the following problem. Let B be a simply- connected domain contained in E and let ζ be a point in B. If $f(z)$ is a function from some set $\mathcal{M}$, what conditions can be imposed on B so that $f(B)$ is starlike with respect to $f(\zeta)$? For example, if $\zeta = 0$ and B is an open disk, then the solution to this problem is just the radius of starlikeness for the set $\mathcal{M}$.

Aleksandrov proves a number of theorems; for example,

Theorem 22. PWO. Each noneuclidean disk B with noneuclidean center ζ and noneuclidean radius $r < \pi/2$ is mapped by an arbitrary $f(z)$ in S onto a domain that is starlike with respect to $f(\zeta)$. The bound $\pi/2$ is sharp.

Sakaguchi 1959 [1-B-1218] considers functions that are starlike with respect to symmetrical points. He calls a function $f(z)$ *starlike with respect to symmetrical points* if for every ϱ in $(1-\epsilon, 1)$ and every ζ with $|\zeta| = \varrho$, the angular velocity of $f(z)$ about the point $f(-\zeta)$ is positive at $z = \zeta$ as z traverses the circle $|z| = \varrho$ in a positive direction. Thus, we have the inequality

$$\text{Re}\,\frac{2\zeta f'(\zeta)}{f(\zeta)-f(-\zeta)} > 0 \tag{83}$$

for all ζ in some ring $1-\epsilon < |\zeta| < 1$, where $\epsilon > 0$.

In Problem 70 we will see that the inequality (83) in $r < |z| < 1$ does not in itself imply univalence, and so we must assume this to obtain the set we wish.

Let STS denote the set of all functions $f(z) = z + \cdots$ that are regular and univalent in E and satisfy the inequality (83) in E.

J. Stankiewicz 1965 [2-B-1385] obtained a structural formula for functions in STS; namely, he proved that $f(z) = z + \cdots$ is in STS if and only if there is a $p(z)$ in P such that

$$f'(z) = p(z)\exp\left(\frac{1}{2}\int_0^z \frac{p(\zeta)+p(-\zeta)-2}{\zeta}\,d\zeta\right).$$

We will look further at the set STS in Problems 71 and 72. For some generalizations of the set STS, see Chand and P. Singh 1979 [3-B-176]. Raymon and Tepper 1975 [2-B-1191] have introduced the concept of the *index of starlikeness* of a function $f(z)$ in ST. This index for a function $f(z)$ is the largest radius δ such that for each ζ in the disk $|z| \le \delta$, the domain $f(E)$ is starlike with respect to $f(\zeta)$. We use ST_δ to denote the subset of ST of functions that have index of starlikeness $\ge \delta$. From this definition we have $ST_1 = CV$ and $ST_0 = ST$. For the set ST_δ, Raymon and Tepper obtain bounds for $|a_n|$, $|f(z)|$, and Re $zf'(z)/f(z)$. For example, they prove that if $f(z)$ is in ST_δ, then for $n \ge 2$,

$$|a_n| \le \prod_{k=1}^{n-1} \frac{k(1+\delta)+1-(-\delta)^k}{k(1+\delta)+\delta+(-\delta)^{k+1}}. \tag{84}$$

Clearly these bounds are sharp when $\delta = 0$ and when $\delta = 1$, but no claim about sharpness is made if $0 < \delta < 1$.

Brannan and Kirwan 1961 [2-B-179] introduce a related concept of $\delta(r)$-visibility. A function $f(z)$ in S is said to be a *$\delta(r)$-visible function* if for some fixed $r < 1$ the domain $f(E_r)$ is starlike with respect to each point w_0 in $|w| \le \delta(r)$.

They prove, among other results, that if $f(z)$ is in S and is $\delta(r)$-visible for $f(E_r)$ with fixed $r < 1$, and $f(z) \not\equiv z$, then for z in E_r

$$|\arg Q_{ST}(z)| \leq \operatorname{Arc} \cos \frac{\delta(r)}{M(r, f)}.$$

Mocanu 1966 [2-B-988] gave a rather complicated definition for a curve $f(C_r)$ to be convex with respect to a curve $f(C_\rho)$ where $0 < \varrho < r \leq 1$. When $\varrho = \delta$ and $r = 1$, the Mocanu definition leads to the same one given by Raymon and Tepper. Mocanu 1969 [2-B-989] used his definition to investigate the particular function $e^z - 1$. For further work see Mocanu 1970 [2-B-991].

5. FURTHER GENERALIZATIONS OF A STARLIKE DOMAIN

In 1954 [1-B-1094] Rahmanov gave six different definitions that represent six variations on the theme of convex and starlike functions. For some further notes on Rahmanov's method, see Avhadiev and Aksent′ev 1975 [2-B-79, p. 24–25]. In this section we follow the notation of Rahmanov and number these new definitions R-1 to R-6. In each definition the domain D lies in the $w = u + iv$ plane, contains the point $w = 0$, and does not contain the point $w = \infty$. In all of these definitions the parameters p, q, and r are assumed to be positive.

Definition R-1. A domain D is said to be of type D_{px} if for each point M of the boundary of D that does not lie on the u-axis, there is a b such that the branch of the parabola $v = \sqrt{2p(u - b)}$ that passes through M, has the property that the piece joining M and the u-axis is contained in D.

Definition R-2. A domain D is said to be of type D_{py} if for each point M of the boundary of D that does not lie on the u-axis, there is a b such that the branch of the parabola $\sqrt{2pv} = b - u$ that passes through M has the property that the piece joining M and the u-axis is contained in D.

In these two definitions p is a constant as indicated by the subscript on D_{px} and D_{py}. Thus the straight lines of ST and CV are replaced by a certain collection of parabolas.

Definition R-3. A domain D is said to be of type $D_{p\theta}$ if for each point M of the boundary of D, there is a θ such that the branch of the parabola $\varrho \sin^2(\psi - \theta) = 2p \cos(\psi - \theta)$ that passes through M has the property that the piece joining M and the origin lies in D. Here (ϱ, ψ) are polar coordinates for the (u, v) plane.

Definition R-4. Let C_r be a fixed circle with radius r and center at the origin and contained in D. We consider a set of rays tangent to the circle C_r at the end point, and we assign a fixed direction (say, counterclockwise) the same for each tangent ray. The domain D is of type D_{rk} if for each boundary point M the segment of the tangent ray between M and the point of tangency on C_r lies in D.

We observe that if $r > 0$, then there is a domain of type D_{rk} that is not starlike with respect to the origin. The letter k in the subscript is suppose to remind us of the tangency property (the Russian work for tangent starts with a k).

Definition R-5. Suppose that a fixed parabola $v^2 = 2pu$ is contained in D. We consider a set of rays each tangent to the parabola at the end point and we assign a fixed direction, the same for each tangent ray. The domain D is said to be of type D_{pk} if for each point M of the boundary of D, the segment of the tangent ray from M to the point of tangency lies in D.

If follows from this definition that a domain of type D_{pk} must be unbounded and (if $p > 0$) it must contain all the points "inside" the parabola (the points for which $2pu > v^2$).

Definition R-6. Suppose that a fixed parabola $v^2 = 2p(u - q)$ lies in the *complement* of D. We consider a set of rays each tangent to the parabola at the end point and we assign a fixed direction, the same for each ray. The domain D is said to be of type D_{dpk} if for each boundary point M of D the segment of the tangent ray from M to the point of tangency lies in the *complement* of D.

Here the subscript d reminds us of the complement (the Russian word for complement starts with a d).

In what follows we assume that if D is a domain of type D_{px}, D_{py}, and D_{pk}, then the boundary of D is a closed Jordan curve in the extended plane that contains the point at ∞.

If D is of type D_{rk} or $D_{p\theta}$, then D may be bounded by a closed Jordan curve that lies in the finite part of the plane.

In the next six theorems we assume that $f(z)$ is regualar in $|z| \leq 1$ except for a finite number of points on the boundary. Further, we always have $f(0) = 0$ and $f'(0) = 1$. With these conditions Rahmanov proves the following six theorems.

Theorem R-1. PWO. The function $f(z)$ is univalent in E and $f(E)$ is a domain of type D_{px} if and only if for $|z| = 1$ we have

$$(\mathrm{Im}\, f(z))(\mathrm{Re}\, zf'(z)) + p\, \mathrm{Im}\, zf'(z) \geq 0, \tag{85}$$

except for possibly a finite number of points.

The next five theorems have exactly the same form but with the replacements indicated.

Theorem R-2. PWO. Replace D_{px} by D_{py} and (85) by

$$\sqrt{2\, \mathrm{Im}\, f(z)}\, \mathrm{Im}\, zf'(z) - \sqrt{p}\, \mathrm{Re}\, zf'(z) \geq 0. \tag{86}$$

Theorem R-3. PWO. Replace D_{px} by $D_{p\theta}$ and (85) by

$$\mathrm{Re}\, [(\sqrt{2}\sqrt{p^2 + |f(z)|^2} + i\sqrt{p\sqrt{p^2 + |f(z)|^2} - p^2})Q_{ST}(z)] \geq 0. \tag{87}$$

Theorem R-4. PWO. Replace D_{px} by D_{rk} and (85) by

$$\mathrm{Re}\left[\left(1 + \frac{ir}{\sqrt{|f(z)|^2 - r^2}}\right) Q_{ST}(z)\right] \geq 0. \tag{88}$$

Theorem R-5. PWO. Replace D_{px} by D_{pk} and (85) by

$$p\, \mathrm{Im}\, zf'(z) + (\mathrm{Im}\, f(z) + \sqrt{[\mathrm{Im}\, f(z)]^2 - 2p\, \mathrm{Re}\, f(z)})\, \mathrm{Re}\, zf'(z) \leq 0. \tag{89}$$

Theorem R-6. PWO. Replace D_{px} by D_{dpk} and (85) by

$$[\mathrm{Im}\, f(z) + \sqrt{[\mathrm{Im}\, f(z)]^2 - 2p\, \mathrm{Re}\, f(z) + 2pq}]\, \mathrm{Re}\, zf'(z) + p\, \mathrm{Im}\, zf'(z) \geq 0. \tag{90}$$

For other domains, see Rahmanov 1953 [1-B-1093].

Noor and Thomas 1980 [3-B-649] introduced the concept of a quasi-convex function. A function $f(z)$, regular in E, with the usual normalization, is said to be *quasi-convex* in E if there is a function $g(z)$ in CV such that

$$\text{(91)} \qquad \operatorname{Re} \frac{(zf'(z))'}{g'(z)} = \operatorname{Re} \frac{f'(z) + zf''(z)}{g'(z)} > 0$$

in E. The set of all such functions will be denoted by QCV.

Observe that if $g(z) \equiv f(z)$, then equation (91) becomes the condition for a convex function. Hence $CV \subset QCV$. The proof that a quasi-convex function is univalent must be postponed until we have covered close-to-convex functions in Chapter 11 and Libera's Lemma (Lemma 15.5). See Problem 15.18.

Taładaj 1970 [3-B-910] has also introduced the term *quasi-convex function* for a function that satisfies a different and far more complicated condition than (91). See Siewierski and Śmiałkówna [2-B-1332] and [3-B-816].

In a convex domain every pair of points can be joined by a line segment contained in the domain. We can modify this definition by allowing the segment to bend slightly but not too much. This idea is made precise by putting an upper bound θ_0 on the variation of arg τ on a curve joining z_1 and z_2 in D. Here τ is the angle that the tangent to the curve makes with the line joining z_1 and z_2. This idea has been developed by Reade 1958 [1-B-1111], Cowling and Royster 1961 [1-B-227], and Sone 1963 [1-B-A-140].

A domain D is said to be of type $\widetilde{R}(\theta_0, d_0)$ if every pair of points in D can be joined in D by a curve Γ of length $\leq d_0$, on which the variation of $\tau \leq \theta_0$. Avhadiev 1974 [2-B-75] has obtained a number of different conditions on $f(z)$ for $f(z)$ to be univalent in a domain of type $\widetilde{R}(\theta_0, d_0)$.

6. FUNCTIONS THAT ARE STARLIKE WITH RESPECT TO A BOUNDARY POINT

In this section we give a brief summary of some of the work by M. S. Robertson in 1981 [3-B-758]. Let $f(z)$ be starlike with respect to a boundary point w_0. By a rotation we may assume that $f(1) = w_0$. We next set

$$\text{(92)} \qquad \phi(z) = \frac{f(z) - w_0}{f(0) - w_0}$$

so that $\phi(0) = 1$ and $\phi(1) = 0$. Further, $\phi(z)$ is starlike with respect to $\phi(1) = 0$. We observe that since $\phi(z)$ is starlike with respect to the origin then $\phi(E)$ must omit some ray and hence we can define $g(z) = \sqrt{\phi(z)}$ so that: (a) $g(z)$ is regular in E, (b) $g(z)$ is univalent and starlike with respect to $g(1) = 0$, (c) $g(0) = 1$, and (d) $g(E)$ is contained in some half-plane whose boundary contains the origin.

Definition 5. Let $RB^\star$ denote the set of all functions

$$g(z) = 1 + \sum_{n=1}^{\infty} d_n z^n \tag{93}$$

that either satisfy the conditions (a), (b), (c), and (d) listed above, or $g(z) \equiv 1$. Further, let $RB_\star$ denote the set of functions in $RB^\star$ that are also regular on ∂E.

Definition 6. Let RB denote the set of functions of the form (93) for which $g(z) \neq 0$ in E and

$$\operatorname{Re}\left(\frac{2zg'(z)}{g(z)} + \frac{1+z}{1-z}\right) \geq 0, \qquad z \text{ in } E. \tag{94}$$

Theorem 23. PWO. For these sets $RB_\star \subset RB \subset RB^\star$.

The inclusion $RB_\star \subset RB^\star$ is trivally true from the definition of these two sets. The difficult part is the proof that $RB \subset RB^\star$. In fact we do not know at present whether $RB = RB^\star$ is true or not.

Notice that if $g(z) \equiv 1$ in E, then (94) is true. This is one reason for adding the constant function to the sets $RB_\star$ and $RB^\star$. We can prove that $RB_\star \subset RB$ as follows. Let $g(z)$ be in $RB_\star$. If $g(z) \equiv 1$, then (94) is true and $g(z)$ is in RB. Suppose $g(z) \not\equiv 1$. By the definition of $RB_\star$ there is an α such that $\operatorname{Re} e^{i\alpha} g(z) > 0$ in E. Therefore, $g(z) \neq 0$ in E. Now $g(1) = 0$ and $g(z)$ is regular at $z = 1$, so the condition that $\operatorname{Re} e^{i\alpha} g(z) > 0$ implies that $g(z)$ has a simple zero at $z = 1$ and $g'(1) \neq 0$. Then $g'(z)/g(z)$ has a simple pole at $z = 1$ with residue 1, and consequently the function

$$p(z) \equiv \frac{2zg'(z)}{g(z)} + \frac{1+z}{1-z} \tag{95}$$

is regular at $z = 1$ (the singularity is removable) and hence regular for all

z in $\bar{E}$. Further $\operatorname{Re} p(z) > 0$ for $|z| = 1$ and $z \neq 1$, because $g(E)$ is starlike with respect to the point $g(1) = 0$. Hence, $\operatorname{Re} p(z) \geq 0$ for all z in $\bar{E}$. Thus $g(z)$ is in RB.

Theorem 24. If $f(z)$ is in $ST(1/2)$, then $g(z) \equiv (1-z)f(z)/z$ is in RB.

Proof. A brief computation gives

$$\frac{2zg'(z)}{g(z)} + \frac{1+z}{1-z} = 2\left(\frac{zf'(z)}{f(z)} - \frac{1}{1-z}\right) + \frac{1+z}{1-z} = 2\frac{zf'(z)}{f(z)} - 1.$$

Therefore, in E

$$\operatorname{Re}\left(\frac{2zg'(z)}{g(z)} + \frac{1+z}{1-z}\right) \geq 2\left(\frac{1}{2}\right) - 1 = 0.$$

Further, if $g(z) = (1-z)f(z)/z$, then $g(0) = 1$ and $g(1) = 0$. ■

The same type of computation will give the converse.

Theorem 25. PLE. Suppose that $f(z) = z + \cdots$ is regular in E and $f(z) = zg(z)/(1-z)$. If $g(z)$ is in RB then $f(z)$ is in $ST(1/2)$.

In Problem 11.29 we will see that if $g(z)$ is in RB, then either $g(z) \equiv 1$ or $g(z)$ is close-to-convex in E, and hence $g(z)$ is univalent in E. When this is established, then it is easy to obtain sharp bounds for the coefficients d_n in the Maclaurin series (93) for $g(z)$ (see Problem 11.35).

PROBLEMS

1. Prove that if $\alpha = 1$, the two sets $CV(\alpha)$ and $ST(\alpha)$ have only the one function $f(z) \equiv z$.
2. With the usual normalization, equation (3), and $0 < \alpha < 1$, prove that $f(z)$ is convex if and only if $g(z) \equiv z[f'(z)]^{\alpha}$ is in $ST(1-\alpha)$.

   Prove that if $f(z)$ is in $ST(1-\beta)$ with $0 < \beta < 1$, then

$$g(z) \equiv z\left(\frac{f(z)}{z}\right)^{\alpha}, \qquad 0 < \alpha < 1,$$

   is in $ST(1-\alpha\beta)$. State and prove the converse. Pinchuk 1971 [2-B-1124, p. 13].

3. Prove that $p(z)$ is in $P(\alpha)$ if and only if there is a $\mu(\phi)$ in $M(0, 2\pi)$ for which

$$p(z) = \alpha + \frac{1-\alpha}{2\pi}\int_0^{2\pi} \frac{1+ze^{-i\phi}}{1-ze^{-i\phi}}\, d\mu(\phi). \tag{96}$$

$$= \frac{1}{2\pi}\int_0^{2\pi} \frac{1+(1-2\alpha)ze^{-i\phi}}{1-ze^{-i\phi}}\, d\mu(\phi).$$

Compare this result with equation (7.80) in Problem 7.37.

4. Prove that $F(z)$ is in $ST(\alpha)$, for $0 \le \alpha < 1$, if and only if there is a $\mu(\phi)$ in $M(0, 2\pi)$ for which $F(z) = ze^{I(z)}$, where

$$I(z) = -\frac{(1-\alpha)}{\pi}\int_0^{2\pi} \ln(1-ze^{-i\phi})\, d\mu(\phi). \tag{97}$$

Show that $I(z)$ can also be written in the form

$$I(z) = 2(1-\alpha)\int_0^z \frac{b(t)}{1-tb(t)}\, dt,$$

where $|b(t)| < 1$ for z in E. Hint: If $Q(z)$ is in P, then $Q(z) = [1+zb(z)]/[1-zb(z)]$ with $|b(z)| < 1$. Schild 1965 [1-B-A-138].

5. Prove the theorems stated in equation (15a) and (15b).
6. Find a Stieltjes integral formula for functions in $CV(\alpha)$.
7. Suppose that $0 < \alpha < 1$ and $f(z)$ has the form (3). Prove that if $\Sigma_2^\infty (n-\alpha)|a_n| \le 1-\alpha$, then $|Q_{ST}(z) - 1| \le 1-\alpha$ and hence $f(z)$ is in $ST(\alpha)$. Merkes, Robertson, and Scott 1962 [1-B-918].
8. State and prove a result similar to the one in Problem 7 for $f(z)$ to be in $CV(\alpha)$.
9. In Problems 7 and 8 are the constants on the right side best possible?
10. Suppose that the infinite product

$$F(z) \equiv z\prod_{k=1}^{\infty} \frac{f_k(z)}{z}$$

converges in E. If $f_k(z)$ is in $ST(1-\alpha_k)$ and $\beta = 1 - \Sigma_{k=1}^\infty \alpha_k \ge 0$, prove that $F(z)$ is in $ST(\beta)$. Merkes, Robertson, and Scott [1-B-918].

11. Observe that $f_k(z) \equiv z - z^3/4n^2$ is starlike of order $1 - 2/(4n^2-1)$. Use this, the result of Problem 10, and the infinite product

$$\frac{2}{\pi}\sin\frac{\pi}{2}z = z\prod_{n=1}^{\infty}\left(1-\frac{z^2}{4n^2}\right)$$

to prove that $\sin(\pi z/2)$ is starlike in E. Robertson 1954 [1-B-1160].

12. The order of convexity and the order of starlikeness do not depend on the normalization $f'(0) = 1$. We require only that $f'(0) \neq 0$. Prove this fact: (a) analytically and (b) geometrically.

13. Let $\partial f(E)$ denote the boundary of the image of E under $f(z)$. A function $f(z)$ may be bounded and starlike, and yet the "curve" $\partial f(E)$ may not be rectifiable (the "length" may be infinite). Fill in the details (very few) in the following argument: Let D be the domain obtained from E by deleting an infinite number of radial slits, $re^{i\theta_n}$, of length 1/2 ending (or starting) on the circle $|w| = 1$ and selected so that θ_n is strictly decreasing to zero. If $g(z)$ maps E onto D, then $g(z)$ is a bounded starlike function for which $\partial g(E)$ has infinite length (is not rectifiable). Eenigenburg 1972 [2-B-352] for Problems 13 and 14.

14. If we add in Problem 13 the condition that $f(z)$ is starlike of order α $(0 < \alpha < 1)$, then $\partial f(E)$ may still be nonrectifiable. Prove this by considering

$$f(z) \equiv z(g(z)/z)^{1-\alpha}$$

where $g(z)$ is the function defined in Problem 13.

15. Suppose that k is a positive integer and $f(z)$ is in $CV(\alpha)$. Prove that

$$g(z) \equiv \int_0^z [f'(t^k)]^{1/k}\, dt$$

is also in $CV(\alpha)$. Since $g(z)$ has k-fold symmetry, it is in $CV^{(k)}(\alpha)$.

16. Suppose that $g(z)$ is in $CV^{(k)}(\alpha)$, for k a positive integer. Prove that

$$f(z) \equiv \int_0^z [g'(t^{1/k})]^k\, dt$$

is in $CV(\alpha)$. Sakaguchi 1973 [2-B-1259].

17. A generalization of Problem 2: Suppose that $f_k(z)$ is in $ST(\alpha_k)$ and $g_k(z)$ is in $CV(\beta_k)$. Let $a_k, b_k > 0$ for each k and set

$$F(z) = z \prod_{k=1}^{m} \left(\frac{f_k(z)}{z}\right)^{a_k} \prod_{k=1}^{n} (g'(z))^{b_k}.$$

Prove that $F(z)$ is in $ST(\gamma)$ where

$$\gamma = 1 - \sum_{k=1}^{m} a_k(1-\alpha_k) - \sum_{k=1}^{n} b_k(1-\beta_k).$$

Note that γ may be negative, and if so, $F(z)$ is not necessarily univalent. If $\gamma \geq 0$, then $F(z)$ is univalent and starlike. Silverman 1975 [3-B-817].

18. Prove that $k^\star(z, \alpha) \equiv z/(1-z)^{2-2\alpha}$ is in $ST(\alpha)$ for any $\alpha < 1$. This function is called the *Koebe function of order* α. Prove that if $\alpha < \beta < 1$, then $k^\star(z, \alpha)$ is not in $ST(\beta)$. If $\epsilon > 0$, find the largest β such that $f(z) \equiv z/(1-z)^\epsilon$ is in $ST(\beta)$. Find the maximum β for $g(z) \equiv z/(1-z^k)^\epsilon$ where k is a positive integer.

19. Use $k^\star(z, \alpha)$ defined in Problem 18 to show that if $\alpha < 0$, then there is at least one function in $ST(\alpha)$ that is not univalent in E.

20. Prove that if k is a positive integer, then $f(z)$ and $[f(z^k)]^{1/k}$ belong to the same set $ST(\alpha)$. Of course, it can happen that neither one has an order of starlikeness (both are in $ST(-\infty)$).

★ **21.** Prove that for each δ in $(0, 1)$ there is a function in $ST(\delta)$ that is not convex. Hint: Use $g(z)$ defined in Problem 18, with k a suitably large integer. See M. Klein 1968 [2-B-712], who obtains this result in a different way.

22. Padmanabhan 1968 [2-B-1083] defines a function to be starlike of order α in E if $f(z)$ has the usual normalization and if for all z in E

$$\left| \frac{Q_{ST}(z) - 1}{Q_{ST}(z) + 1} \right| < \alpha, \qquad 0 < \alpha \leq 1.$$

He obtains quite a few results for this new class of starlike functions. Prove that $f(z)$ is in this new class if and only if there is a $b(z)$ with $|b(z)| \leq \alpha$ in E such that $f(z) = ze^{I(z)}$ where

$$I(z) = 2\int_0^z \frac{b(t)}{1 - tb(t)}\, dt.$$

Compare this $I(z)$ with the one in Problem 4.

23. Suppose that $f(z)$ is in ST and that for fixed A and B with $-1 \leq B < A \leq 1$ we have $Q_{ST} \prec (1 + Az)/(1 + Bz)$. Find a structural formula for the set of all $f(z)$ that satisfy these conditions. Show that this formula gives the formulas in Problems 4 and 22 as special cases.

24. Use the results obtained in Problem 7.11 to prove the coefficient bounds stated in Theorem 2. Hint: Use the technique of dominant power series, Chapter 7.3.

25. Let $P^{(k)}(\alpha)$ be the set of functions

$$p(z) = 1 + \sum_{n=1}^{\infty} b_{nk} z^{nk}$$

which are regular in E, and in E satisfy $\operatorname{Re} p(z) > \alpha$, with $\alpha < 1$. Find the sharp upper bounds for $|b_{nk}|$.

26. Find sharp upper bounds for the coefficients of

$$g(z) = z + \sum_{n=1}^{\infty} b_{nk+1} z^{nk+1}$$

if (a) $g(z)$ is in $ST^{(k)}(\alpha)$ and (b) $g(z)$ is in $CV^{(k)}(\alpha)$, where $0 \leq \alpha < 1$ and k is a positive integer. For a generalization to functions of the form $z + \Sigma_{k+1}^{\infty} a_n z^n$ see Boyd 1966 [2-B-170].

27. Suppose that $\alpha \neq 0$. Are there any fixed points for the transformation (33)?

28. Suppose that $\alpha = -1/N$ where N is a positive integer. Show that the transformation (35) may give a function that is not regular in E. Hint: Consider the starlike function $F(z) = z/(1 + cz)$, where $0 < |c| < 1$. Mocanu 1969 [2-B-990] for Problems 28 through 30.

29. With $F(z)$ as in Problem 28, find $f(z)$ when (a) $\alpha = 1$, (b) $\alpha = 1/2$, and (c) $\alpha = 2$.

30. Using (35) find $f(z)$ when $F(z) = z/(1 + z)^2$ and $\alpha = 1/2$.

31. Prove that if $f(z)$ is in $\alpha - CV$, then

$$\ln \left\{ \left(\frac{f(z)}{z} \right)^{1-\alpha} [f'(z)]^{\alpha} \right\} = -\frac{1}{\pi} \int_0^{2\pi} \ln (1 - ze^{-i\phi})\, d\mu(\phi),$$

where $\mu(\phi)$ is in $M(0, 2\pi)$. Mocanu and Reade 1971 [2-B-999].

32. Suppose that $f(z)$ has the form of equation (3) and $g(z)$ has the form given in Problem 26, and both are regular in E. Prove that $f(z)$ is in $k\alpha - CV$ if and only if $g(z) \equiv f(z^k)^{1/k}$ is in $\alpha - CV$, where k is a positive integer. Coonce and S. Miller 1974 [2-B-291].

★ **33.** Suppose that $\lambda_k > 0$ for each k and $\Sigma_{k=1}^{n} \lambda_k = 2$. Set

$$f(z) = \frac{z}{\prod_{k=1}^{n} (1 - \epsilon_k z)^{\lambda_k}},$$

where $|\epsilon_k| = 1$, and the ϵ_k are distinct. Recall that $f(z)$ maps E onto the complex plane cut along n radial slits making angles $\pi\lambda_1, \ldots,$ $\pi\lambda_n$ at the origin. Set $\beta = \min \{\lambda_k\}$. Prove that $f(z)$ is in $(-\alpha) - CV$ for each α in $[0, \beta]$. Eenigenburg 1974 [2-B-353].

34. What is the greatest α such that the Koebe function is in $\alpha - CV$?

35. Set $F(z) \equiv \dfrac{2zf'(z)}{f(z)-f(\zeta)} - \dfrac{z+\zeta}{z-\zeta}$ for fixed ζ in E. Prove that if $f(z)$ is univalent in E then $F(z)$ has a removable singularity at $z = \zeta$ and hence is regular in E.

36. Continuation of Problem 35: Prove that

$$\lim_{z\to\zeta} F(z) = 1 + \frac{zf''(\zeta)}{f'(\zeta)} \equiv Q_{CV}(\zeta).$$

37. If in Problems 35 and 36 we let $\zeta \to z$, do we get $Q_{CV}(z)$?

38. Suppose that in Problem 35, $f(z)$ is in CV. Let $z = re^{i\theta}$ and $\zeta = re^{i\alpha}$ where α is a constant. Observe that $\arg[f(re^{i\theta}) - f(re^{i\alpha})]$ is nondecreasing for $\alpha \le \theta \le \alpha + 2\pi$. Prove that

$$2\frac{\partial}{\partial\theta}(\arg[f(re^{i\theta}) - f(re^{i\alpha})]) = \operatorname{Re}\frac{2zf'(z)}{f(z)-f(\zeta)} \ge 0.$$

But under these conditions $\operatorname{Re}[(z+\zeta)/(z-\zeta)] = 0$. Hence for all pairs z, ζ in E, we have $\operatorname{Re} F(z) \ge 0$. Sheil-Small 1969 [2-B-1319] and Suffridge 1970 [2-B-1405].

39. Using the results of Problems 35 through 38, prove the Marx-Strohhäcker Theorem: If $f(z)$ is convex in E and $f(0) = 0$, then $f(z)$ is starlike of order $1/2$ in E. Is $1/2$ the best possible constant? Marx 1932 [1-B-909] and Strohhäcker [1-B-1340]. For a different proof of this theorem, see Komatu 1961 [1-B-723]. For a beautiful generalization of the Marx-Strohhäcker Theorem see Suffridge 1976 [3-B-897, Theorem 10].

40. Assume that $|z| < |\zeta|$ and $|f(z)| < |f(\zeta)|$, and expand $F(z)$ from Problem 35 in a Maclaurin series. Show that the coefficient of z is $2/\zeta - 2/f(\zeta)$. What theorems follow from this? Marx [1-B-909] and Strohhäcker [1-B-1340].

41. Suppose that $f(z) = z + \Sigma_{n=2}^{\infty} a_n z^n$ and has no other zeros except the one at the origin ($f(z)$ is not assumed to be univalent). Prove that if $\operatorname{Re}\sqrt{f(z)/z} > 1/2$, then $|a_n| \le n$ for all n. Dvorák 1933 [1-B-268].

Let R be the largest number such that $\operatorname{Re}\sqrt{f(z)/z} > 1/2$ for all $|z| < R$ and all $f(z)$ in S. In the same paper Dvorák proved that $R > 0.83$. In [2-B-338 and 339] he tried to prove that $R > 0.97$. Unfortunately, this work contains an error, as pointed out simultaneously by Duren and Schober 1971 [2-B-336], Reade and

Umezawa [2-B-1200], and Kuhnau [2-B-767]. Each of these authors finds the precise value of R as the root of a transcendental equation. Reade finds that $R = 0.83559\ldots$. Goel 1975 [2-B-458] has extended these results to functions that satisfy Re $\sqrt{f(z)/z} > 1/2\alpha$, where $\alpha \geq 1$.

42. Derive equation (43) from equation (42).

43. Prove Theorem 9.

44. If we write $|a_n| \leq B_n$ for (47) and $|a_n| \leq b_n$ for (48), we have proved that $|b_n| \leq B_n$. Prove this inequality directly from the expressions for b_n and B_n.

45. Suppose that $0 < \varrho < 1$, $0 \leq \alpha \leq \pi/2$, and $s = e^{-i\alpha}\cos\alpha$. Prove that $f(z) = z/(1-z)^{2s(1-\rho)}$ is in $SP(\alpha, \varrho)$.

46. Prove that with the usual normalization

$$f(z) \equiv z\left(\frac{F(z)}{z}\right)^s, \qquad s = e^{-i\alpha}\cos\alpha,$$

is in $SP(\alpha)$ if and only if $F(z)$ is in ST.

47. Prove that $f(z) = i(1-z)^i - i = z + \cdots$ is in $CVSP(\alpha)$. What value of α must be used for this function?

48. Prove that the function defined in Problem 47 has infinitely many zeros in E. See Krzyz and Lewandowski 1963 [1-B-A-80] and Robertson 1969 [2-B-1221].

49. Let $f(z) = \dfrac{1}{t}\left(\dfrac{1}{(1-z)^t} - 1\right)$ where t may be complex. For what values of t is $f(z)$ in $CVSP(\alpha)$ for some α? Robertson [2-B-1221].

50. Find necessary and sufficient conditions on t for $f(z)$, defined in Problem 49, to be univalent in E. Hint: Use the periodic properties of $e^w = e^{-t\ln(1-z)}$. Royster 1965 [2-B-1231].

51. Combine Problems 49 and 50 to prove Robertson's result, namely if $1/2 < \cos\alpha < 1$, there is an $f(z)$ in $CVSP(\alpha)$ which is not univalent in E.

52. Prove that if $n \geq 2$ is an integer, then $f(z) \equiv z + z^2 + \cdots + z^n$ is *not* univalent in E. Compare this with Theorems 12, 13, and 14.

★ **53.** Suppose that $f(z)$ is in CV and $|\alpha| \leq \pi/2$. Prove that $\Phi(z) \equiv f(z) + e^{i\alpha}zf'(z)$ is univalent in E. Hint: Use a geometric argument to prove that the image of C_r is a simple closed curve. Rahmanov 1951 [1-B-1090].

54. Prove that if $f(z)$ is univalent and starlike with respect to symmetrical points (see equation (83)), then $|a_n| \leq 1$ and this upper bound is sharp for each $n \geq 2$. Sakaguchi 1959 [1-B-1218].

55. If $f(z)$ is in $ST(\beta)$ with $0 < \beta \leq 1$, prove that $g(z) \equiv f(z)/(1-z)^{2\beta}$ is in ST. For $\beta = 1/2$ see Robertson 1981 [3-B-758, p. 332].

56. Prove Theorem 25.

57. Prove that if $f(z)$ is a normalized quasi-convex function, then we have $|a_n| \leq 1$ for all n. Hence $|f(z)| \leq r/(1-r)$ and $|f'(z)| \leq 1/(1-r)^2$. Hint: Use the technique of dominant power series. Noor and Thomas 1980 [3-B-649].

58. Prove that if $\Sigma_1^\infty n|a_n| \leq 1$, then $\phi(\zeta) \equiv \zeta + \Sigma_1^\infty a_n/\zeta^n$ is starlike in E^x, that is, $\phi(\zeta)$ maps E^x onto a domain for which the complement is starlike with respect to the origin.[1] See Problem 8.6.

59. Prove that if $\Sigma_1^\infty n^2|a_n| \leq 1$, then $\phi(\zeta)$, from Problem 58, is convex in E^x, that is, maps E^x onto a domain for which the complement is a convex set.

60. Let Γ be a curve in the $w = Re^{i\Phi}$-plane directed by its parametrization $R = R(t)$ and $\Phi = \Phi(t)$. Let β be the directed angle from Γ to the counterclockwise directed circle $|w| = R$ at a point of intersection of Γ and the circle. Here $-\pi < \beta \leq \pi$. Prove that

$$\tan \beta = \frac{R'(t)}{R(t)\Phi'(t)}, \tag{98}$$

with suitable agreements if $\Phi'(t) = 0$.

61. The graph Γ_L of $R = Ae^{B\Phi}$, $A > 0$, B real, is called a logarithmic spiral (or equiangular spiral). Let β_L denote the angle β of Problem 60 for Γ_L. Prove that $\tan \beta_L = B$ when Φ is selected as the parameter t. Thus β_L is the same at every point of intersection of Γ_L with a circle $|w| = R$.

62. Let $\delta = \arg Q_{ST}(f(z))$ where $\delta = 0$ when $z = 0$. Prove that $\alpha - \pi/2 < \delta < \alpha + \pi/2$, if $f(z)$ is in $SP(-\alpha)$.

★ **63.** Let $\Gamma_\theta = f(C_r)$ and let β_θ denote the angle of Problem 60 for Γ_θ and the circle $|w| = R$. Prove that $\tan \beta_\theta = -\tan \delta$ where δ is defined in Problem 62.

64. Use a continuity argument with z near 0 to prove that $\beta_\theta = -\delta$. Hence $-\alpha - \pi/2 < \beta_\theta < -\alpha + \pi/2$.

[1] The complement may have no interior points. If the complement consists of a finite number of radial line segments, then it is a starlike set.

65. Assume that $\alpha > 0$ and $f(z)$ is in $SP(-\alpha)$. In the logarithmic spiral of Problem 61 set $B = \tan(\pi/2 - \alpha)$ so that $\beta_L = \pi/2 - \alpha > \beta_\theta$. This proves that as R decreases, Γ_L crosses Γ_θ from the "outside" to the "inside." Remember that $f(z)$ is univalent and $\Gamma_\theta = f(C_r)$ is a simple closed curve.

66. Let $f(z)$ be in $SP(-\alpha)$ with $\alpha > 0$. Suppose that P_0 is an arbitrary point in $f(E)$ and Γ_L is the spiral through P_0 defined in Problems 61 and 65. Prove that the bounded component of Γ_L (with one end at P_0) is contained in $f(E)$. This accounts for the name "spirallike" for functions in $SP(\alpha)$.

67. Give a suitable interpretation of Problems 62, 65, and 66 when $\alpha = 0$.

68. Remove the restriction $\alpha > 0$ made in Problems 65 and 66.

69. Prove that $\widetilde{ST}(\lambda) = SP(\alpha) \cap SP(-\alpha)$, if $0 \leq \alpha < 1$ and $\lambda = 1 - \alpha$. Hence any strongly starlike function is also spirallike for a suitable value of the parameter λ. For a deeper analysis of the geometric properties presented in Problems 60 through 69, see J. Stankiewicz 1966 [2-B-1383] and 1969 [2-B-1386].

70. Find a function regular in E that satisfies the inequality (83) in some ring domain $r < |\zeta| < 1$ and is not univalent in E.

71. Suppose that $f(z)$ is in S and that (83) holds in some ring domain $r < |\zeta| < 1$. Prove that then (83) holds throughout E and hence $f(z)$ is in STS. For a partial converse see Problem 11.26.

★ **72.** Let D be a domain, and z_1 and z_2 any two points in D. Let $L(z_1, z_2)$ be the length of a smooth curve in D that joins z_1 and z_2. We consider the ratio $L(z_1, z_2)/|z_2 - z_1|$ and we say that a domain D is of type λ if

$$\lambda \equiv \sup_{z_1, z_2 \in D} \frac{\inf L(z_1, z_2)}{|z_2 - z_1|}$$

is finite. Suppose that $F(z)$ is regular in D and not a constant. Prove that if there is a c such that for all z in a domain D of type λ, $|F'(z) - c| < |c|/\lambda$, then $F(z)$ is univalent in D. See Rogozin [1].

73. For the domain E^x, find λ from Problem 72.

74. Suppose that $F(\zeta) = \zeta + \Sigma_{n=0}^{\infty} a_n/\zeta^n$ is regular in E^x and that $|F'(\zeta) - 1| \leq 1$ in E^x. Prove that either $F(\zeta)$ is univalent in $\bar{E}^x$ or $F(\zeta) = \zeta + a_0 + e^{i\alpha}/\zeta$. Notice that this result does not follow from

Problems 72 and 73. Compare this result with Theorems 17.1 and 17.2. Aksent′ev 1958 [1-B-22]. See also Avhadiev and Aksent′ev 1975 [2-B-79, p. 13].

75. Prove that if $f(z)$ is in $CV(\alpha)$, then $|\arg f'(z)| \leq 2(1-\alpha)\,\text{Arc}\sin r$, and this result is sharp. This is the Rotation Theorem for $CV(\alpha)$. Pinchuk 1968 [2-B-1122].

CHAPTER 10

TYPICALLY-REAL FUNCTIONS AND RELATED TOPICS

1. UNIVALENT FUNCTIONS WITH REAL COEFFICIENTS

As usual let

$$f(z) = z + \sum_{n=2}^{\infty} a_n z^n \tag{1}$$

be univalent in E. The central conjecture that $|a_n| \leq n$ is still an open problem; but as we saw in Chatper 8, if $f(E)$ is starlike, then it is true that $|a_n| \leq n$ for all n. It is natural to search for other subsets of S for which $|a_n| \leq n$ can be proved. One such easily accessible subset is given in

Definition 1. Let SR denote the subset of functions in S for which all of the coefficients a_n are real.

In general we adopt the convention that the addition of the letter R denotes the subset of functions for which all the coefficients are real.

Thus *CVR* is the set of all functions that are normalized convex univalent and have all coefficients real in the power series expansion (1).

Almost simultaneously, Dieudonné 1931 [1-B-242, p. 318], Rogosinski 1932 [1-B-1180, pp. 97 and 116], and Szász 1932 [1-B-1349] proved

Theorem 1. If $f(z)$ is in SR, then for $n = 2, 3, 4, \ldots$

$$-n \leq a_n \leq n. \tag{2}$$

Proof. Szász [1-B-1349]. Set $z = re^{i\theta}$ and consider the difference quotient

$$\begin{aligned}\Delta(re^{i\theta}) &\equiv \frac{f(z)-f(\bar{z})}{z-\bar{z}} = 1 + \sum_{n=2}^{\infty} \frac{a_n r^n(e^{in\theta}-e^{-in\theta})}{r(e^{i\theta}-e^{-i\theta})} \\ &\equiv 1 + \sum_{n=2}^{\infty} a_n r^{n-1} \frac{\sin n\theta}{\sin \theta}\end{aligned} \tag{3}$$

for $0 < r < 1$ and $0 < \theta < \pi$. We extend the definition of Δ to include $\theta = 0$ and $\theta = \pi$ by agreeing that $\Delta(r) = f'(r)$ and $\Delta(-r) = f'(-r)$. We note that

$$\lim_{\theta \to 0} \frac{\sin n\theta}{\sin \theta} = n, \qquad \lim_{\theta \to \pi} \frac{\sin n\theta}{\sin \theta} = (-1)^{n+1} n. \tag{4}$$

If $f(z)$ is univalent, then $\Delta(re^{i\theta}) \neq 0$ for the region $0 \leq r < 1$, and $0 \leq \theta \leq 2\pi$. Further, $\Delta(0) = 1$. Since all the coefficients are real, $\Delta(z)$ is real and hence $\Delta(z) > 0$ in E.

Now set

$$I_n \equiv \int_0^{\pi} \Delta(re^{i\theta}) \sin \theta \sin n\theta \, d\theta. \tag{5}$$

Then, by the orthogonality of $\{\sin k\theta\}$ on $[0, \pi]$,

$$\begin{aligned} I_n &= \int_0^{\pi} \left(1 + \sum_{k=2}^{\infty} a_k r^{k-1} \frac{\sin k\theta}{\sin \theta}\right) \sin \theta \sin n\theta \, d\theta \\ &= a_n r^{n-1} \int_0^{\pi} \sin^2 n\theta \, d\theta. \end{aligned} \tag{6}$$

On the other hand

$$I_n = \int_0^{\pi} \Delta(re^{i\theta}) \sin \theta \sin n\theta \, d\theta = \int_0^{\pi} \Delta(re^{i\theta}) \sin^2\theta \frac{\sin n\theta}{\sin \theta} \, d\theta.$$

If m_n and M_n denote the minimum and maximum values of $\sin n\theta/\sin\theta$ in $[0, \pi]$, then

$$m_n \int_0^\pi \Delta(re^{i\theta}) \sin^2\theta \, d\theta \le I_n \le M_n \int_0^\pi \Delta(re^{i\theta}) \sin^2\theta \, d\theta.$$

Using the series expansion (3) for $\Delta(re^{i\theta})$ and the orthogonality of $\{\sin k\theta\}$, we obtain

$$m_n \int_0^\pi \sin^2\theta \, d\theta \le I_n \le M_n \int_0^\pi \sin^2\theta \, d\theta. \tag{7}$$

Finally, we combine (6) and (7) and note that both trigonometric integrals give $\pi/2$. This gives

$$m_n \le a_n r^{n-1} \le M_n. \tag{8}$$

Now $M_n = n$, and $m_n \ge -n$. But min $(\sin n\theta/\sin\theta) = -n$ only if n is even. If n is odd, then $m_n > -n$. If we let $r \to 1$ in (8), then we obtain $-n \le m_n \le a_n \le n$. ■

We note in passing that we have proved more than was announced in Theorem 1. Indeed, we have

Theorem 2. If $f(z)$ is in SR and $m_n \equiv \min(\sin n\theta/\sin\theta)$, then for each n

$$m_n \le a_n \le n \tag{9}$$

and this inequality is sharp for each n.

Proof. Of course, the Koebe function shows that the upper bound is sharp. The function

$$\tilde{k}_\phi(z) \equiv \frac{z}{1 - 2z\cos\phi + z^2} = z + \sum_{n=2}^\infty \frac{\sin n\phi}{\sin\phi} z^n \tag{10}$$

is in SR, and $\tilde{k}_\phi(z)$ shows that the lower bound is sharp. We merely select ϕ so that $\sin n\theta/\sin\theta$ has its minimum at $\theta = \phi$. ■

We also observe that in the proof of Theorem 1 we did not need the univalence of $f(z)$. We needed only the fact that $f(z) \ne f(\bar{z})$ for all z in the upper half of the unit disk. Thus $f(z)$ might map E onto a domain of the type shown in Figure 1, and the coefficient bounds stated in Theorem 2 would still be true. This remark suggests the study of the wider class of functions called typically-real functions. This class was introduced in 1932 by Rogosinski [1-B-1180].

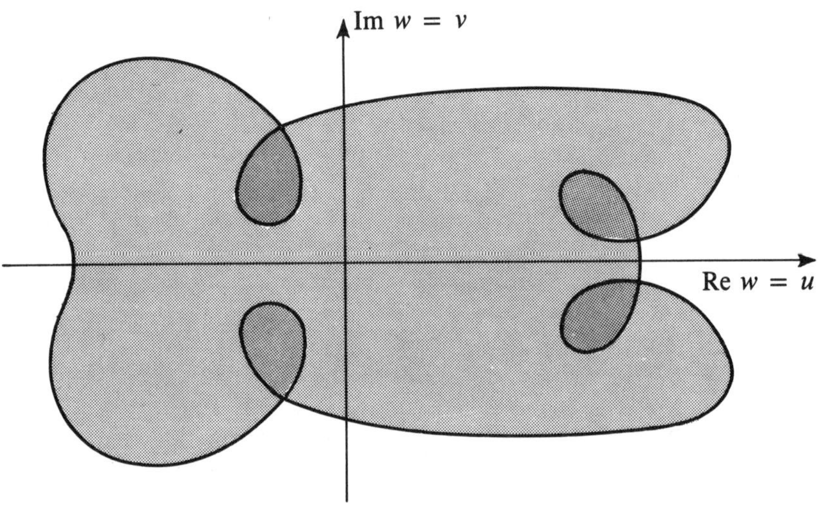

FIGURE 1.

2. TYPICALLY-REAL FUNCTIONS

Definition 2. If the function

$$f(z) = z + \sum_{n=2}^{\infty} a_n z^n \tag{1}$$

is regular in E and if for every nonreal z in E

$$\text{sign}\,(\text{Im}\, f(z)) = \text{sign}\,(\text{Im}\, z), \tag{11}$$

then $f(z)$ is said to be *typically-real* in E. The class of all such functions is denoted by TR.

The condition (11) is often written in the equivalent form

$$(\text{Im}\, f(z))(\text{Im}\, z) \geq 0. \tag{12}$$

This definition can be modified to include other normalizations, domains different from E, and meromorphic functions. One merely preserves the condition (11) which is the central concept for a typically-real function.

The proofs of Theorems 1 and 2 give

Theorem 3. Rogosinski [1-B-1180]. If $f(z)$ is in TR, then the coefficients satisfy the inequality (9) and this inequality is sharp for each n.

Typically-real functions are related to functions with a positive real part in a surprisingly simple way.

Definition 3. Let PR denote the set of functions

$$g(z) = 1 + \sum_{n=1}^{\infty} p_n z^n \tag{13}$$

in P for which all the coefficients p_n are real.

Theorem 4. Rogosinski [1-B-1180]. There is a one-to-one correspondence between the two sets TR and PR. If $g(z)$ is in PR, then

$$f(z) \equiv \frac{z}{1-z^2} g(z) = z + p_1 z^2 + (p_2 + 1)z^3 + \cdots \tag{14}$$

is in TR. Conversely, if $f(z)$ is in TR, then

$$g(z) \equiv \frac{1-z^2}{z} f(z) = 1 + a_2 z + (a_3 - 1)z^2 + \cdots \tag{15}$$

is in PR.

Proof. We first prove that if $f(z)$ is in TR, then $g(z)$ is in PR. The series (15) shows that $g(z)$ is regular in E (the pole of $(1 - z^2)/z$ at $z = 0$ is cancelled by the zero of $f(z)$ at the same point). Since the coefficients of $f(z)$ are real, so also are the coefficients of $g(z)$. Further, $g(0) = 1$.

Next assume that $f(z)$ is regular on ∂E and set $z = e^{i\theta}$. Then

$$\text{Re}\, g(z) = \text{Re}\left(\frac{1-z^2}{z} f(z)\right) = \text{Re}\,(-2i \sin \theta f(z)).$$

$$\text{Re}\, g(z) = 2 \sin \theta\, \text{Im}\, f(z) = 2(\text{Im}\, z)(\text{Im}\, f(z)) \geq 0. \tag{16}$$

Since $\text{Re}\, g(z)$ is a harmonic function, the inequality (16) which holds on the boundary of E can be extended to all of E. If $f(z)$ is not regular on the boundary of E, we consider $f(rz)/r$ and let $r \to 1^-$. Then $g(z)$ is in PR.

Conversely, suppose that $g(z)$ is in PR and $f(z) \equiv zg(z)/(1-z^2)$. The same type of argument used before can be applied to the two semi-disks $E^+ \equiv E \cap \{\mathrm{Im}\, z > 0\}$ and $E^- \equiv E \cap \{\mathrm{Im}\, z < 0\}$ to prove that $(\mathrm{Im}\, f(z))(\mathrm{Im}\, z) \geq 0$ in E. This is the argument originally used by Rogosinski [1-B-1180, p. 113]. However, a little more care is needed because of the poles of $z/(1-z^2)$ at ± 1. Here we select a different proof. We recall Theorem 7.25, which we restate here as

Theorem 5. A function $g(z)$ is in PR if and only if it can be written in the form

$$g(z) \equiv \frac{1}{2\pi}\int_0^{2\pi} \frac{1-z^2}{1-2z\cos\phi + z^2}\, d\mu(\phi), \tag{17}$$

where $\mu(\phi)$ is in $M(0, 2\pi)$.

If $g(z)$ is in PR, and $f(z)$ is defined by (14), then the representation (17) for functions in PR gives

$$f(z) = \frac{1}{2\pi}\int_0^{2\pi} \frac{z}{1-2z\cos\phi + z^2}\, d\mu(\phi). \tag{18}$$

We observe that the integrand

$$\frac{z}{1-2z\cos\phi + z^2} = \tilde{k}_\phi(z) = z + \cdots \tag{19}$$

is univalent in E (see Problems 2 through 5). Further, $\tilde{k}_\phi(z)$ has real coefficients and the proper normalization, and hence $\tilde{k}_\phi(z)$ is in TR.

Finally, it is reasonably clear that the set TR is a convex set (see Problem 7). Thus, by a limiting process the integral (18) is in TR. ■

If we start with a function $f(z)$ in TR, use (15) to obtain a function $g(z)$ in PR, and then use equations (17) and (18) to return to $f(z)$, we obtain

Theorem 6. Robertson 1935 [1-B-1140]. The function $f(z)$ is in TR if and only if it can be written in the form (18) for some $\mu(\phi)$ in $M(0, 2\pi)$.

Since $\cos(-\phi) = \cos\phi$, the integrals in (17) and (18) can be replaced by

(20) $$g(z) = \frac{1}{\pi}\int_0^\pi \frac{1-z^2}{1-2z\cos\phi+z^2}\,d\mu(\phi), \qquad g(z)\in PR,$$

and

(21) $$f(z) = \frac{1}{\pi}\int_0^\pi \frac{z}{1-2z\cos\phi+z^2}\,d\mu(\phi), \qquad f(z)\in TR,$$

where now

(22) $$\int_0^\pi d\mu(\phi) = \pi.$$

In equations (20) and (21) the function $\mu(\phi)$ is not necessarily identical with the $\mu(\phi)$ in (17) and (18) in the interval $(0, \pi)$. Frequently in the literature, the factors $1/2\pi$ and $1/\pi$ are dropped in equations (17), (18), (20), and (21), and $\mu(\phi)$ is called a probability distribution or a mass distribution over the circle (or the half-circle). In this case we replace π by 1 on the right side of (22).

The two integral representations, (17) and (18), provide tools for proving in an easy way, many theorems that might otherwise be difficult. For example, if $f(z) = z + \Sigma_{n=2}^\infty a_n z^n$ is in TR, then (18) and (10) give

(23) $$\begin{aligned} f(z) &= \frac{1}{2\pi}\int_0^{2\pi}\left(\sum_{n=1}^\infty \frac{\sin n\phi}{\sin\phi} z^n\right) d\mu(\phi) \\ &= \sum_{n=1}^\infty \left(\frac{1}{2\pi}\int_0^{2\pi} \frac{\sin n\phi}{\sin\phi}\, d\mu(\phi)\right) z^n, \end{aligned}$$

and hence for $n = 1, 2, 3, \ldots$

(24) $$a_n = \frac{1}{2\pi}\int_0^{2\pi} \frac{\sin n\phi}{\sin\phi}\, d\mu(\phi).$$

From equation (24) it is a trivial matter to deduce the sharp bounds for the coefficients that were given in Theorem 3.

Further, since

(25) $$\sin\theta + \sin 3\theta + \sin 5\theta + \cdots + \sin(2n-1)\theta = \frac{\sin^2 n\theta}{\sin\theta},$$

equation (24) gives

$$1 + a_3 + a_5 + \cdots + a_{2n-1} = \frac{1}{2\pi}\int_0^{2\pi} \frac{\sin^2 n\phi}{\sin^2\phi}\, d\mu(\phi).$$

Thus, if $f(z)$ is in TR, then

$$\text{(26)} \qquad 1 + a_3 + a_5 + \cdots + a_{2n-1} \geq 0.$$

If $g(z)$ is in PR, then so is $1/g(z)$. Let $f(z)$ be in TR. Then by Theorem 4

$$g(z) \equiv \frac{1-z^2}{z} f(z)$$

is in PR. Consequently,

$$\frac{1}{g(z)} = \frac{z}{(1-z^2)f(z)}$$

is in PR, and, using Theorem 4 again, we obtain

Theorem 7. Rogosinski 1932 [1-B-1180]. If $f(z)$ is in TR, then so also is

$$\text{(27)} \qquad f_1(z) \equiv \frac{z^2}{(1-z^2)^2 f(z)}.$$

3. THE REGION OF VALUES AT A POINT

Let z_0 be a fixed point and let $\mathcal{M}$ be a set of functions, each regular at z_0. As $f(z)$ runs through all of the functions of the set $\mathcal{M}$, $w = f(z_0)$ generates a set of points called the *region of values of the set $\mathcal{M}$ at z_0*. We can denote this set by $\mathcal{R}(\mathcal{M}(z_0))$, but sometimes we shall abuse this elaborate notation by dropping $\mathcal{M}$ if the set of functions is obvious and by writing $\mathcal{R}(z)$ or $\mathcal{R}(\zeta)$ (in place of z_0) when it is clear that z or ζ are fixed.

As a simple example, let $\mathcal{M}$ be the set of functions that satisfy the conditions of Schwarz's Lemma. Then $\mathcal{R}(z_0)$ is just the closed disk $|w| \leq r = |z_0|$.

The determination of $\mathcal{R}(\mathcal{M}(z_0))$ may be difficult, but if there is an integral representation for functions in $\mathcal{M}$, then it may be relatively easy to describe $\mathscr{R}(\mathscr{M}(z_0))$. We suppose that $f(z)$ is in $\mathscr{M}$ if and only if it can be represented in the form

$$f(z) = \frac{1}{2\pi}\int_0^{2\pi} K(z, \phi)\, d\mu(\phi), \tag{28}$$

where the *kernel* $K(z, \phi)$ is analytic in z and continuous in ϕ and $\mu(\phi)$ is in $M(0, 2\pi)$. For example, if $\mathscr{M}$ is the set P, then

$$K(z, \phi) = \frac{1 + ze^{-i\phi}}{1 - ze^{-i\phi}}, \tag{29}$$

and equation (28) is just the Herglotz formula, equation (7.61).

Theorems 5 and 6 give the two kernels:

$$K(z, \phi) = \frac{1 - z^2}{1 - 2z\cos\phi + z^2} \tag{30}$$

for the set PR, and

$$K(z, \phi) = \frac{z}{1 - 2z\cos\phi + z^2} \tag{31}$$

for the set TR.

As ϕ runs through the interval $[0, 2\pi]$ the kernel generates a curve Γ in the w-plane (recall that z is fixed). Since the integral in (28) is just a weighted average of the points on Γ, we have immediately

Theorem 8. Suppose that the functions in $\mathscr{M}$ have an integral representation of the form (28), where $K(z, \phi)$ is analytic for z in E and continuous in ϕ for ϕ in $[0, 2\pi]$. Then for each fixed z in E the region of values for the set $\mathscr{M}$ is the convex cover of the curve Γ: $w = K(z, \phi)$, $0 \le \phi \le 2\pi$. This region is a bounded closed convex set.

The concepts involved in Theorem 8 were well known to many mathematicians and were frequently used prior to 1955. However, it seems that Ašnevic and Ulina 1955 [1-B-60] were the first to express the result in this general form and to exploit Theorem 8 in a systematic way. We give here only a few of the theorems that can be derived from this result.

We apply Theorem 8 to the set TR of typically-real functions. Since the image of E is symmetric about the real axis, it is sufficient to consider only points in E for which $\operatorname{Im}(z) \geq 0$. Assume first that $\operatorname{Im}(z) > 0$ and examine

$$(32)\qquad w \equiv \frac{1}{K(z,\phi)} = \frac{1+z^2-2z\cos\phi}{z} = z+\frac{1}{z}-2\cos\phi,$$

where $K(z, \phi)$ is the kernel for typically-real functions (see Theorem 6). For fixed z and $0 \leq \phi \leq 2\pi$, the right side of (32) describes a segment of a line that is parallel to the real axis. The distance of this line from the real axis is $|\operatorname{Im}(z + 1/z)|$. Then the curve Γ is a part of the circle obtained when we consider $1/w$. The circle has diameter end points at 0 and $|\operatorname{Im}(z+1/z)|^{-1}$. Γ is that arc of the circle that does not contain 0 and has endpoints at $w_1 = z/(1-z)^2$ and $w_2 = z/(1+z)^2$. The radius of the circle is $1/2|\operatorname{Im}(z+1/z)|$.

Theorem 9. Let $f(z)$ be in TR and suppose that $\operatorname{Im}(z) > 0$. Then the region of values of TR at z is the convex cover of Γ described above. If $z = r$ is real, then the region[1] of values is the line segment joining $r/(1-r)^2$ and $r/(1+r)^2$.

In Figure 2 the shaded region is the region of values for the set TR when $z = (1+i)/2$.

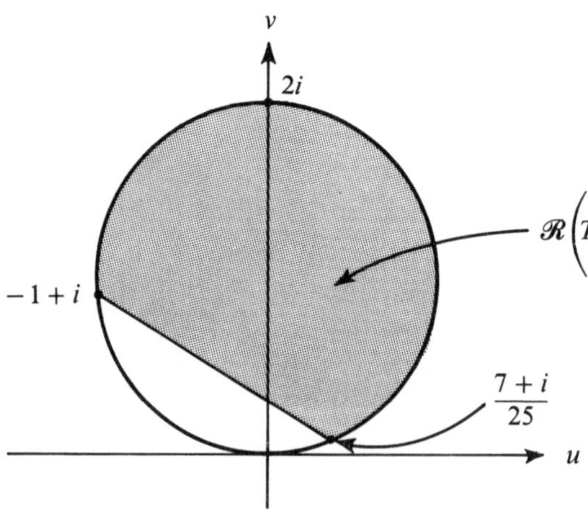

FIGURE 2.

[1] Of course, a line segment is not a region, but it represents the set of possible values for $f(z)$. We prefer to keep the work "region" in this special case.

As a corollary we have

Theorem 10. PLE. Goluzin 1950 [1-B-413]. Let $f(z)$ be in TR and suppose that $\operatorname{Im}(z) > 0$ and z is in E. Then

$$(33)\qquad |f(z)| \le \begin{cases} \left|\dfrac{z}{(1-z)^2}\right|, & \text{if } \operatorname{Re}\left(\dfrac{1+z^2}{z}\right) \ge 2, \\[2ex] \dfrac{1}{\left|\operatorname{Im}\left(\dfrac{1+z^2}{z}\right)\right|}, & \text{if } \left|\operatorname{Re}\dfrac{1+z^2}{z}\right| \le 2, \\[2ex] \left|\dfrac{z}{(1+z)^2}\right|, & \text{if } \operatorname{Re}\left(\dfrac{1+z^2}{z}\right) \le -2. \end{cases}$$

Further,

$$(34)\qquad \arg \frac{z}{(1+z)^2} \le \arg f(z) \le \arg \frac{z}{(1-z)^2}.$$

Remizova 1963 [1-B-1125] used Theorem 9 to obtain the sharp lower bounds for $|f(z)|$ when $f(z)$ is in TR.

Since we can differentiate (22), we find that if $f(z)$ is in TR, then

$$(35)\qquad f'(z) = \frac{1}{2\pi}\int_0^{2\pi} \frac{1-z^2}{(1-2z\cos\phi + z^2)^2}\, d\mu(\phi),$$

with similar but more complicated formulas for the higher derivatives $f''(z)$, $f'''(z)$, etc. Consequently, Theorem 8 gives the exact region of values for each derivative of a typically-real function at an arbitrary point in E. We observe that the kernel (31) can be put in the form

$$(36)\qquad K(z,\phi) = \frac{1}{2i\sin\phi}\left(\frac{1}{1-ze^{i\phi}} - \frac{1}{1-ze^{-i\phi}}\right)$$

and hence for each $n \ge 1$

$$(37)\qquad K^{(n)}(z,\phi) = \frac{n!}{2i\sin\phi}\left(\frac{e^{in\phi}}{(1-ze^{i\phi})^{n+1}} - \frac{e^{-in\phi}}{(1-ze^{-i\phi})^{n+1}}\right).$$

However, for large values of n it may be difficult to describe the curve Γ in any useful way. Sharp upper and lower bounds for $|f'(z)|$ (the case $n = 1$) were obtained by Remizova [1-B-1125].

Following Rogosinski 1932 [1-B-1180], we can use Theorem 4 to transfer the result of Theorem 9 back to the class PR. Then the transformation $b(z) = (g(z) - 1)/(g(z) + 1)$ will give us a useful theorem about the set BR_0. Indeed we multiply each point of the domain described in Theorem 9 by $(1 - z^2)/z$ and obtian

Theorem 11. PLE. Let z be a fixed point in E and let Γ be that arc of the circle through 0, $(1 + z)/(1 - z)$, and $(1 - z)/(1 + z)$ that does not contain the point zero. Then the region of values for the set PR at z is the convex cover of Γ.

This set is shown in Figure 3 for $z = (1 + i)/4$.

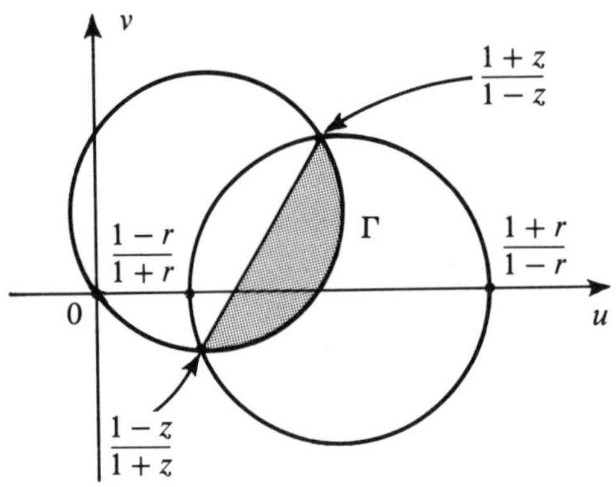

FIGURE 3.

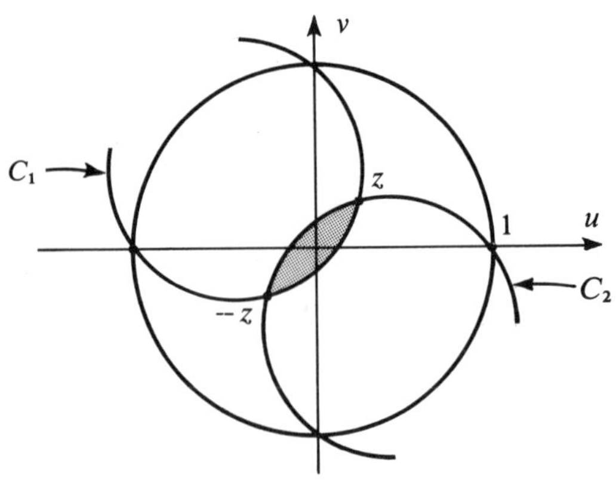

FIGURE 4.

Theorem 12. PLE. Let $f(z)$ satisfy the condition of Schwarz's Lemma and further suppose that the coefficients are all real (the set BR_0). Let z be a fixed point in E, let C_1 be the circle through -1, $-z$, and z, and let C_2 be the circle through the points $-z$, z, and 1. Then the region of values for the set BR_0 at z is that region bounded by the arcs of C_1 and C_2 that contain z and $-z$ but not -1 or 1.

This region is shown in Figure 4 for $z = (1 + i)/3$.

Suppose now that in place of considering the full set TR we select a subset by setting the second coefficient a_2 equal to some fixed constant A_2 where of course $-2 \leq A_2 \leq 2$. Then each one of our earlier theorems presents a new problem. Such a problem is sometimes referred to as a *Gronwall type problem*, because he was the first to suggest such a study in connection with the set S; see 1916 [1-B-454] and 1920 [1-B-455]. Subsets of TR with fixed second coefficient have been studied by Jenkins 1961 [1-B-626], Goluzina 1962 [1-B-427], and Leeman 1973 [2-B-819]. Goluzina found sharp upper and lower bounds as functions of A_2 for each of the four quantities: $|f(z)|$, $|\arg f(z)|$, $|\operatorname{Im} f(z)|$, and $|\operatorname{Re} f(z)|$. Leeman obtained the sharp coefficient bound for $}a_n|$, as a function of A_2, for each $n \geq 3$.

Typically-real polynomials have been treated by Royster and Suffridge 1970 [2-B-1237]] and Merkes and Wright 1971 [2-B-929].

For bounded typically-real functions, the reader should consult the work of Lewandowski and Wajler 1974 [3-B-534].

4. FUNCTIONS THAT ARE CONVEX IN ONE DIRECTION

We say that a domain D is convex in the direction of a line L if each line parallel to L (and including L) either misses D, or is contained in D, or the intersection with D is either a segment or a ray[1] (see Figure 5 next page). Clearly a domain of this type does not need to be convex or starlike with respect to any point.

Definition 4. If $f(z)$ maps E onto a domain convex in the direction of L, we say that $f(z)$ is *convex in the direction of* L. The set of all func-

[1] Briefly, the intersection is either empty or a connected set.

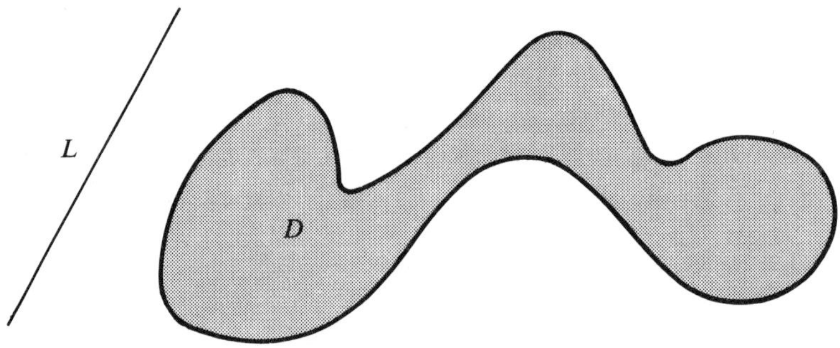

FIGURE 5.

tions $f(z)$ regular in E, with $f(0) = 0$ and $f'(0) = 1$, that are convex in the direction of L will be denoted by $CV(L)$. When L is the imaginary axis, the set will be denoted by $CV(i)$.

Of course, if $g(z)$ is in $CV(L)$, then a suitable rotation $f(z) = e^{i\gamma}g(e^{-i\gamma}z)$ gives a new function in $CV(i)$, so we concentrate our attention on the set $CV(i)$.

A more efficient definition of a domain that is convex in the direction of the i-axis is this: Whenever w_1 and w_2 are in D and $\operatorname{Re} w_1 = \operatorname{Re} w_2$, then all of the points $tw_1 + (1-t)w_2$ with $0 < t < 1$ are also in D.

Fejér 1933 [1-B-295, p. 61] was the first to call attention to this interesting set $CV(i)$, but Robertson 1936 [1-B-1142] and [1-B-1144] was the first to make an intensive study of $CV(i)$.

To begin with, let us make the simplifying assumption that $f(z)$ in $CV(i)$ is also regular in $|z| \leq 1$. If we write $f(e^{i\theta}) = u(\theta) + iv(\theta)$, then (see Figure 6) it is clear that there is some arc, $\alpha < \theta < \beta$, where $u(\theta)$ is decreasing, and on the complementary arc, $u(\theta)$ is increasing.

Since $f(z)$ is regular on $|z| = 1$, we can write

$$\textbf{(38)} \qquad \frac{\partial}{\partial\theta}u(\theta) = \frac{\partial}{\partial\theta}\operatorname{Re} f(z) = \operatorname{Re}\frac{\partial}{\partial\theta}f(z)$$

$$= \operatorname{Re} f'(z)\frac{\partial z}{\partial\theta} = \operatorname{Re} izf'(z) = -\operatorname{Im} zf'(z).$$

It follows that under the conditions described, we have

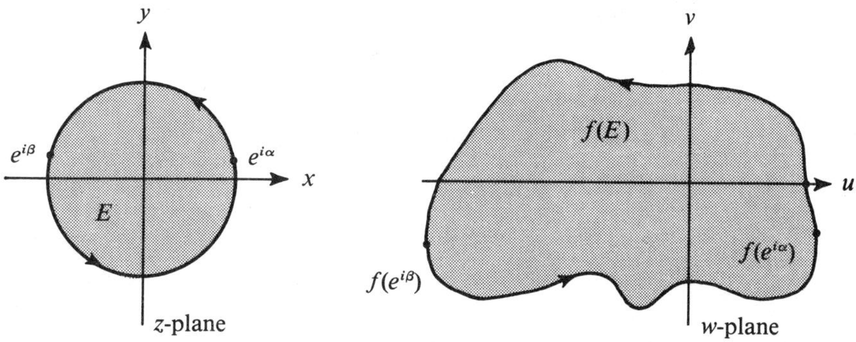

FIGURE 6.

$$
(39) \qquad \begin{cases} -u'(\theta) = \operatorname{Im} zf'(z) \geq 0, & \text{for } \alpha \leq \theta \leq \beta, \\ -u'(\theta) = \operatorname{Im} zf'(z) \leq 0, & \text{for the complementary arc.} \end{cases}
$$

It is natural to ask: if $f(z)$ is in $CV(i)$, does this imply that for each r in $(0, 1)$ the function $f(rz)/r$ is also in $CV(i)$. One is tempted to observe that $\operatorname{Im} zf'(z)$ is a harmonic function and to apply some minimum principle. But this must fail because the harmonic function changes sign in E. To the contrary (as we will soon prove) it may be false that $f(rz)/r$ is also in $CV(i)$. To rescue a theorem from this confused situation we assume that the coefficients are all real. Then we have

Theorem 13. Fejér and Szegö 1951 [1-B-301]. If $f(z)$ is regular in $\bar{E}$, has all coefficients real, has the usual normalization, and satisfies the inequalities (39) on the boundary $z = e^{i\theta}$, then $f(z)$ is univalent in E and $f(rz)/r$ is in $CV(i)$ for each r in $(0, 1]$.

Proof. Since all the coefficients are real, we have $\operatorname{Im} zf'(z) = 0$ for $-1 \leq z \leq 1$. It follows that in (39) we have $\alpha = 0$ and $\beta = \pi$. The minimum principle for harmonic functions can now be applied to $\bar{E}^+$, the closed upper half of the unit disk, and we have $\operatorname{Im} zf'(z) \geq 0$ in $\bar{E}^+$.

By symmetry in the real axis it is sufficient to consider only E^+. To settle the univalence of $f(z)$ we use the mixed Cauchy-Riemann equations (1.32) and find that in E^+

$$
(40) \qquad \frac{\partial v}{\partial r} = -\frac{1}{r}\frac{\partial u}{\partial \theta} = \frac{1}{r} \operatorname{Im} zf'(z) > 0
$$

so that v is increasing on each radial line from the origin. Hence $f(z)$ is typically-real. The condition $u_\theta < 0$, now true on each half-circle $|z| = r, 0 < \theta < \pi$ shows that $f(z)$ is univalent in E^+. ■

Corollary. Under the conditions of Theorem 13, $zf'(z)$ is in TR.

We return to the set $CV(i)$, where $f(z)$ is not necessarily regular for $z = e^{i\theta}$, and the coefficients may be complex. A function in $CV(i)$ is univalent by definition because $f(E)$ is a domain convex in the direction of the i-axis.

It was a great surprise to the mathematical community to learn that if $f(z)$ is in $CV(i)$, this *does not imply* that $f(rz)/r$ is also in $CV(i)$. This was first observed by Hengartner and Schober 1973 [3-B-357], but their proof used an existence argument rather than an example function.

A concrete example that is more informative was given by Goodman and Saff 1979 [3-B-304]. Let

$$F(z) = \frac{z - Az^2}{(1 - Bz)^2}, \qquad A = e^{2i\alpha} \cos \alpha, \qquad B = e^{i\alpha}, \tag{41}$$

where α is real and $0 < \alpha < \pi$.

With a little effort (see Problem 40) the reader can prove that $F(z)$ maps E onto the entire complex plane minus a vertical slit that starts at

$$\begin{aligned} s_0 = F(e^{-3i\alpha}) &= -\frac{\cos \alpha}{2} + i\left(\frac{\sin \alpha}{2} - \frac{1}{4 \sin \alpha}\right) \\ &= \frac{-ie^{-2i\alpha}}{4 \sin \alpha}, \end{aligned} \tag{42}$$

and extends upward to $-(\cos \alpha)/2 + i\infty$. Consequently, $F(z)$ is in $CV(i)$. To investigate the shape of the image of the level curves $|z| = r$ it is simpler to look at $G(z) \equiv F(e^{-i\alpha}z)$. A moderate computation gives for $G(z)$

$$\begin{aligned} \frac{\partial u}{\partial \theta} &\equiv u'(\theta) = -\operatorname{Im} zG'(z) \\ &= \frac{-r(1 - r^2)[(1 + r^2) \sin(\theta - \alpha) + r(3 \sin \alpha - \sin(2\theta + \alpha))]}{|1 - z|^6} \end{aligned} \tag{43}$$

If $r = 1$ we find that $u'(\theta) = 0$ except for the pole at $z = 1$, and this is consistent with the description of $F(E)$, which as a domain is the same as $G(E)$.

However, with a little effort one can see that for some $r_0 < 1$, the right side of (42) has four changes of sign for $0 < \theta < 2\pi$, if α is selected properly (see Problems 41 and 42). Thus, the property that $f(E)$ is convex in the direction of the i-axis is not always preserved for smaller disks.

A more elaborate analysis can be made to prove that in this example we can use any $r_0 > \sqrt{2} - 1 \approx 0.4142$.

In the other direction Umezawa 1952 [1-B-1428] has proved that if $f(z)$ is in S then $f(rz)/r$ is in $CV(i)$ for every $r < 4 - \sqrt{13} \approx 0.3945$. Thus we have

Theorem 14. Let R be the radius of convexity in the direction of the i-axis for the set $CV(i)$. Then $4 - \sqrt{13} \leq R \leq \sqrt{2} - 1$.

The exact value of this R is unknown.

The example function (41) also solves (negatively) another problem mentioned in Chapter 9. Recall that $ST(\zeta)$ is the set of normalized functions for which $f(E)$ is starlike with respect to $f(\zeta)$, and that if $f(E_r)$ is a domain that is starlike with respect to $f(\zeta)$ for $|\zeta| = \varrho < r < 1$, then for $|z| = r$ we have

$$\text{(44)} \qquad \operatorname{Re} \frac{zf'(z)}{f(z) - f(\zeta)} \geq 0.$$

If the condition (44) is satisfied for $|z| = 1$, can it be extended into an annular domain $1 - \epsilon < |z| < 1$? The answer is no. The function $F(z)$ defined by (41) is starlike with respect to each point w_0 that lies on the vertical slit and below the point $w = s_0$. But if $F(rz)$ does not carry E onto a domain convex in the direction of the i-axis (as we have proved), then $F(E_r)$ is certainly not starlike with respect to w_0 for any r in $(\sqrt{2} - 1, 1)$.

The presence of examples such as $F(z)$ makes the analysis of the set $CV(i)$ rather difficult. To resolve this difficulty Hengartner and Schober 1970 [2-B-571] discard the usual normalization and adopt one that is more tractable. We leave $f(0)$ unspecified (although no harm is done if we set $f(0) = 0$) and we impose no condition on $f'(0)$ except that $f'(0) \neq 0$.

The following analysis and the next five theorems are due to Hengartner and Schober. We will not give the complete details, but we will outline their work. The industrious reader may fill in the details or consult the original paper.

Suppose that $f(z)$ is regular in E and $f(E) = D$ is a domain that is convex in the direction of the i-axis. Now D may be bounded or unbounded from the left or right, but we can always set

$$m = \inf_{z \in E} \operatorname{Re} f(z), \qquad \text{and} \qquad M = \sup_{z \in E} \operatorname{Re} f(z), \tag{45}$$

with the reasonable agreement that we may have $m = -\infty$ and/or $M = \infty$. In any case we can find an infinite sequence of points z_n' in E such that

$$\lim_{n \to \infty} \operatorname{Re} f(z_n') = M \tag{46}$$

and a second sequence of points z_n'' in E such that

$$\lim_{n \to \infty} \operatorname{Re} f(z_n'') = m. \tag{47}$$

From each of these infinite sequences we can select a convergent subsequence, which by the definition of m and M must converge to a point on ∂E. Suppose that $z_n' \to e^{i\alpha}$ and $z_n'' \to e^{i\beta}$. If there are several possible limits (for example, D might be a rectangle), then we try to select our subsequences so that $\alpha \neq \beta$. Now such a selection is not always possible. Consider the two example functions:

$$f_1(z) = \frac{z}{1 - iz} \qquad \text{and} \qquad f_2(z) = \frac{z}{(1 - iz)^2}. \tag{48}$$

Both functions map E onto domains convex in the direction of the i-axis: $f_1(E)$ is the half-plane $\operatorname{Im} w < 1/2$, and $f_2(E)$ is the complex plane minus the slit $w = ti$, $t \geq 1/4$. For both of these functions the sequences z_n' and z_n'' converge to the same point, the pole at $z = -i$, so that for these functions (and others as well) we have $\alpha = \beta = 3\pi/2$.

We first consider the case in which we can select subsequences such that $\alpha \neq \beta$. By a Möbius transformation $L(z)$ of E we can carry 1 into $e^{i\alpha}$ and -1 into $e^{i\beta}$. Thus we set $g(z) = f(L(z))$. Then the sequences z_n' and z_n'' now defined for $g(z)$ will converge to 1 and -1, respectively. This is the normaliztion selected by Hengartner and Schober. Summarizing, with $g(z)$ replaced by $f(z)$, we have

Definition 5. We let $\widetilde{CV}_2(i)$ be the set of all functions $f(z)$ regular in E such that:

(a) $f(E)$ is convex in the i direction,
(b) there is a sequence z'_n of points in E such that

$$\lim_{n\to\infty} z'_n = 1 \quad \text{and} \quad \lim_{n\to\infty} \operatorname{Re} f(z'_n) = \sup_{z\in E} \operatorname{Re} f(z) \tag{49}$$

and

(c) there is a sequence z''_n of points in E such that

$$\lim_{n\to\infty} z''_n = -1 \quad \text{and} \quad \lim_{n\to\infty} \operatorname{Re} f(z''_n) = \inf_{z\in E} \operatorname{Re} f(z). \tag{50}$$

When can we normalize a function so that it is in $\widetilde{CV}_2(i)$? It is easy to prove that if $f_1(z)$ is in $CV(i)$ and the complement of $f_1(E) = D$ contains two vertical half-lines, one of which meets ∂D from above, and the other meets ∂D from below, then there is an $L(z)$ such that $f(z) \equiv f_1(L(z))$ is in $\widetilde{CV}_2(i)$.

For brevity we use the symbol $D_2(i)$ to denote the set of domains of the type just described. Thus, D is in $D_2(i)$ if and only if it is convex in the direction of the i-axis and the complement of D contains two vertical half-lines that meet ∂D, one from above and one from below. With these elaborate preparations we have the main theorem.

Theorem 15. PWO. Suppose that $f(z)$ is regular and not a constant in E. Then we have

$$\operatorname{Re}(1 - z^2)f'(z) \geq 0, \qquad z \in E, \tag{51}$$

if and only if the following three conditions are met:
(a) $f(z)$ is univalent in E,
(b) $f(E)$ is in $D_2(i)$,
(c) $f(z)$ has the normaliztion dictated by (49) and (50).

Briefly we can say that $\operatorname{Re}(1 - z^2)f'(z) \geq 0$ is a necessary and sufficient condition for $f(z)$ to be in $\widetilde{CV}_2(i)$.

We first give a crude argument that explains the factor $1 - z^2$ in (51). We consider only those points on C: $|z| = 1$ for which $f'(z)$ exists. Then for $z = e^{i\theta}$

$$\operatorname{Re}(1 - z^2)f'(z) = \operatorname{Re} \frac{1 - z^2}{z} zf'(z) = \operatorname{Re}(-2i \sin\theta\, zf'(z))$$

$$= \operatorname{Im}(2 \sin\theta\, zf'(z)).$$

Thus, if (51) is satisfied on C, then $u'(\theta) \leq 0$ for the upper half of the unit circle and $u'(\theta) \geq 0$ for the lower half of the unit circle (see the inequalities (39)). This certainly suggests that $f(E)$ is in $D_2(i)$.

To give a more suitable argument let γ_t be an arc of a circle that joins $z = -1$ to $z = 1$ in E. Each such arc can be parametrized by

$$z = z(s) = \frac{e^{s+it} - 1}{e^{s+it} + 1}, \qquad -\infty < s < \infty, \tag{52}$$

for each t in $(-\pi/2, \pi/2)$. The value $t = 0$ gives the straight line segment from -1 to 1. A short computation shows that on γ_t (going from -1 to 1)

$$\frac{d}{ds} \operatorname{Re}\,(f(z(s))) = 2 \operatorname{Re}\,\{[1 - z^2(s)]\,f'(z(s))\} \geq 0, \tag{53}$$

and hence $\operatorname{Re} f(z)$ is increasing or constant on each γ_t.

Consider now the family of circular arcs $\{\delta_s\}$ in E that are orthogonal to the family $\{\gamma_t\}$. If we have strict inequality in (51), then by the conformality of $f(z)$, we have $\operatorname{Im} f(z)$ is increasing on each δ_s. If z_1 and z_2 are any two points in E, there is a path joining z_1 and z_2 that consists of at most one subarc of a γ_t and one subarc of a δ_s. Hence $f(z_1) \neq f(z_2)$ and $f(z)$ is univalent in E. Further, if (51) holds with strict inequality, the behavior of $\operatorname{Re} f(z)$ on each γ_t shows that the image of the domain bounded by $\gamma_\tau \cup \gamma_t \cup \{1, -1\}$ with $-\pi/2 < \tau < t < \pi/2$ is of type $D_2(i)$ and, further, $f(z)$ has the normalization described by (49) and (50). If we let $\tau \to -\pi/2$ and $t \to \pi/2$, we see that the condition $\operatorname{Re}\,(1 - z^2)f'(z) > 0$ in E is sufficient to insure properties (a), (b), and (c) of Theorem 15, and $f(z)$ is in $\widetilde{CV}(i)$.

If the equal sign in (51) holds at just one point in E, then it holds throughout E. Then it is easy to prove that

$$f(z) = a_0 + i\,\frac{c_1}{2} \ln \frac{1+z}{1-z}, \qquad c_1 \text{ real}, \tag{54}$$

and this function maps E onto a vertical strip.

The proof of the converse part of Theorem 15 is more delicate and we omit it.

Theorem 16. PLE. If

$$f(z) = a_0 + (b_1 + ic_1)z + \sum_{n=2}^{\infty} a_n z^n, \qquad b_1, c_1 \text{ real}, \tag{55}$$

is regular in E and satisfies $\text{Re}\,(1-z^2)f'(z) \geq 0$ in E, then for $n \geq 1$

$$|a_{2n}| \leq b_1 \tag{56}$$

and

$$|a_{2n+1}| \leq \left(1 - \frac{1}{2n+1}\right) b_1 + \frac{1}{2n+1} |b_1 + ic_1|. \tag{57}$$

and hence $|a_n| \leq |f'(0)|$ for every $n \geq 2$.

Observe that the condition $\text{Re}\,(1-z^2)f'(z) \geq 0$ implies that $b_1 \geq 0$. If $b_1 = 0$, then $\text{Re}\,(1-z^2)f'(z) = 0$ throughout E, and hence $f(z)$ has the form (54). In this case

$$f(z) = a_0 + ic_1 \sum_{n=0}^{\infty} \frac{z^{2n+1}}{2n+1}, \tag{58}$$

and the even subscripted coefficients are all zero. Further, with $b_1 = 0$, the inequalities (57) are sharp for each n. If $c_1 = 0$, then the function

$$f(z) = \frac{b_1}{1-z} \tag{59}$$

shows that both (56) and (57) are sharp for each n. Finally, if $b_1 c_1 \neq 0$, then the inequality (57) may not be sharp but it comes very close, as the function

$$f(z) = \frac{b_1}{1-z} + i\frac{c_1}{2}\ln\frac{1+z}{1-z}, \qquad b_1 > 0, \tag{60}$$
$$= \sum_{n=0}^{\infty} b_1 z^n + i \sum_{n=0}^{\infty} \frac{c_1 z^{2n+1}}{2n+1}$$

shows.

Theorem 17. PLE. Under the conditions of Theorem 16,

$$\frac{(1-r)|f'(0)|}{(1+r)(1+r^2)} \leq |f'(z)| \leq \frac{|f'(0)|}{(1-r)^2} \tag{61}$$

and

$$\frac{1}{2}|f'(0)|\ln\frac{(1+r)^2}{1+r^2} \leq |f(z)-f(0)| \leq \frac{r|f'(0)|}{1-r}. \tag{62}$$

We now examine the exceptional domains that were excluded in Definition 5.

First, let us recall that Liouville's Theorem guarantees that for a regular univalent function, $f(E)$ cannot be the entire complex plane and hence some point w_0 must be a boundary point of $f(E)$. Then $f(E)$ must either omit the vertical half-line $w_0 + it \text{ Re } (w_0)$, $t \geq 0$, or it must omit the vertical half-line with $t \leq 0$. Of course, $f(E)$ may omit many such vertical half-lines all directed upward (or all directed downward) and still not be a domain of type $D_2(i)$. This leads to

Definition 6. A domain D (not the entire plane) is in the set $D_1^+(i)$ if whenever $u_0 + iv_0$ is in D, then all the points $u_0 + i(v_0 + t)$, $t \geq 0$, are also in D.

The domain D is in the set $D_1^-(i)$ if whenever $u_0 + iv_0$ is in D, then all the points $u_0 + i(v_0 - t)$, $t \geq 0$, are also in D.

For example, $f_1(E)$ and $f_2(E)$ are in $D_1^-(i)$, where $f_1(z)$ and $f_2(z)$ are defined by equation (48).

We should observe that the sets $D_2(i)$, $D_1^+(i)$, and $D_1^-(i)$ are not mutually exclusive. A vertical strip is in all three of these sets.

Suppose that $f(z)$ maps E onto a domain in $D_1^+(i)$ or $D_1^-(i)$. Then some point on the boundary of E is the preimage of the point at infinity, and by a Möbius transformation $L(z)$ we can let this point be $z = 1$. In this case we can find two sequences z_n' and z_n'' of points in E such that:

(63) $$\lim_{n\to\infty} z_n' = 1 \quad \text{and} \quad \lim_{n\to\infty} f(z_n') = \sup_{z\in E} \text{Re } f(z)$$

and

(64) $$\lim_{n\to\infty} z_n'' = 1 \quad \text{and} \quad \lim_{n\to\infty} f(z_n'') = \inf_{z\in E} \text{Re } f(z)$$

Theorem 18. PWO. Suppose that $f(z)$ is regular and not a constant in E. Then we have

(65) $$\text{Im } (1 - z)^2 f'(z) \geq 0, \qquad z \in E,$$

if and only if the following three conditions are met:

(a) $f(z)$ is univalent in E,

(b) $F(E)$ is in $D_1^+(i)$,

(c) $f(z)$ has the normalization dictated by (63) and (64).

If we apply Theorem 18 to $-f(z)$, we obtain

Theorem 19. Suppose that $f(z)$ is regular and not a constant in E. Then we have

$$\text{(66)} \qquad \operatorname{Im} (1-z)^2 f'(z) \le 0, \qquad z \text{ in } E.$$

if and only if the following three conditions are met:

(a) $f(z)$ is univalent in E,

(b) $f(E)$ is in $D_1^-(i)$,

(c) $f(z)$ has the normalization dictated by (63) and (64).

Since $f(z) = \pm iz/(1-z)^2$ satisfies the conditions of Theorems 18 and 19, it is impossible to prove any better bounds than $|a_n| \le n|a_1|$ from the inequality (65) or (66).

Theorem 20. Robertson 1936 [1-B-1142]. Suppose that $f(z)$ is regular in E, $f(0) = 0$, and $f'(0) = 1$. Then $f(E)$ is a domain convex in the direction of the i-axis if and only if there are real numbers μ, ν in $[0, \pi]$ such that in E

$$\text{(67)} \qquad \operatorname{Re} \{ -ie^{i\mu}(1 - 2e^{-i\mu} \cos \nu \, z + e^{-2i\mu} z^2) f'(z)\} \ge 0.$$

Robertson does not state Theorem 20 in the form given here, but it is clearly implied by his work on functions starlike in one direction. The condition (67) is given by Royster and Ziegler in 1976 [3-B-762].

The Robertson proof predates by thirty-four years the counter-example given by Hengartner and Schober. Robertson avoids this difficulty by assuming that $f(z)$ satisfies at least one of two conditions:

(a) $f(z)$ is regular on $|z| = 1$ and $\operatorname{Im} zf'(z)$ has exactly two changes of sign there,

(b) there is a $\delta > 0$ such that for each r in $(1-\delta, 1)$, $\operatorname{Im} zf'(z)$ has exactly two changes of sign on $|z| = r$.

Now the function defined in (41) maps E onto a domain convex in the i direction and satisfies neither (a) nor (b). Thus by the nature of the situation Robertson's proof cannot cover the entire set $CV(i)$. However, if we start from the Hengartner and Schober Theorems (15, 18, and 19), then we can deduce Robertson's condition (67) for *every* function in $CV(i)$ as Royster and Ziegler showed in [3-B-762].

Proof of Theorem 20. If $f(E)$ is a domain in the set $D_2(i)$, then there is a point $z = e^{i\alpha}$ that corresponds to sup Re $(f(z))$ and a second point $z = e^{i\beta}$ that corresponds to inf Re $(f(z))$ for z in E. (See the discussion connected with equations (45), (46), and (47).) There is a Möbius transformation

$$\textbf{(68)}\quad \zeta = L(z) = e^{i\gamma}\frac{z-a}{1-\bar{a}z} \qquad \text{or} \qquad z = M(\zeta) = \frac{\zeta e^{-i\gamma}+a}{1+\bar{a}e^{-i\gamma}\zeta}$$

that carries $z = 1$ and -1 into $\zeta = e^{i\alpha}$ and $e^{i\beta}$, respectively. Hence, $g(z) \equiv f(L(z))$ is in $\widetilde{CV}_2(i)$, and we can apply Theorem 15 to $g(z)$. Now

$$(1-z^2)g'(z) = (1-z^2)f'(L(z))e^{i\gamma}\frac{1-|a|^2}{(1-\bar{a}z)^2}.$$

Since $L(z)$ carries E onto E, we can replace $L(z)$ by ζ and z by $M(\zeta)$. A brief computation with these replacements gives

$$\begin{aligned}(1-z^2)g'(z) &= \frac{(1-|a|^2)f'(\zeta)e^{i\gamma}[(1+\bar{a}e^{-i\gamma}\zeta)^2-(e^{-i\gamma}\zeta+a)^2]}{(1+\bar{a}e^{-i\gamma}\zeta-\bar{a}e^{-i\gamma}\zeta-|a|^2)^2}\\ &= \frac{e^{i\gamma}[1-a^2-2e^{-i\gamma}(a-\bar{a})\zeta-(1-\bar{a}^2)e^{-2i\gamma}\zeta^2]f'(\zeta)}{1-|a|^2}.\end{aligned}$$

Therefore, Re $(1-z^2)g'(z) \geq 0$ in E if and only if

$$\textbf{(69)}\qquad \operatorname{Re}\,(1-a^2)e^{i\gamma}\left[1-\frac{2e^{-i\gamma}(a-\bar{a})}{1-a^2}\zeta-\frac{1-\bar{a}^2}{1-a^2}e^{-2i\gamma}\zeta^2\right]f'(\zeta) \geq 0$$

for ζ in E. Now set $1-a^2 = \varrho e^{i\delta}$ so that $1-\bar{a}^2 = \varrho e^{-i\delta}$. A brief computation shows that if $|a| < 1$, then $|(a-\bar{a})/(1-a^2)| < 1$, and hence we can set

$$\frac{a-\bar{a}}{1-a^2} = e^{-i\delta}i\cos\nu$$

for some ν in $[0, \pi]$. Finally, if we set $\mu = \gamma+\delta+\pi/2$, then (69) will give (67). Since (67) is true at $z = 0$, we have Re $(-ie^{i\mu}) = \sin\mu \geq 0$ or $0 \leq \mu < \pi$.

Suppose that $f(E)$ is a domain in the set $D_1^+(i)$. If the two sequences described in equations (63) and (64) converge to $e^{i\mu}$, we set

$$g(z) \equiv f(e^{i\mu}z)$$

and $\zeta = e^{i\mu}z$. Then $g(z)$ satisfies the conditions of Theorem 18, and consequently,

$$\begin{aligned}\operatorname{Im}(1-z)^2 g'(z) &= \operatorname{Im}(1-z)^2 e^{i\mu} f'(e^{i\mu} z) \geq 0 \\ &= \operatorname{Im} e^{i\mu}(1-e^{-i\mu}\zeta)^2 f'(\zeta) \geq 0 \\ &= \operatorname{Re}\{-ie^{i\mu}(1-e^{-i\mu}\zeta)^2 f'(\zeta)\} \geq 0\end{aligned}$$

for ζ in E. This is identical with (67) when $\nu = 0$.

Finally, suppose that $f(E)$ is a domain in $D_1^-(i)$. This time we let $-e^{i\mu}$ be the distinguished value of z and set

$$g(z) = f(-e^{i\mu}z), \qquad \zeta = -e^{i\mu}z.$$

Then by Theorem 19

$$\operatorname{Im}(1-z)^2 g'(z) = \operatorname{Im}\{-(1-z)^2 e^{i\mu} f'(-e^{i\mu}z)\} \leq 0$$

or

$$\operatorname{Im}(1-z)^2 e^{i\mu} f'(-e^{i\mu}z) \geq 0$$

$$\operatorname{Re}\{-ie^{i\mu}(1+e^{-i\mu}\zeta)^2 f'(\zeta)\} \geq 0$$

and this is (67) when $\nu = \pi$. ■

Once we have a formula such as (67) for a set of functions, it is easy to prove theorems for the set. Some of these theorems will be given as problems. Observe that in equation (67) the value $z = 0$ gives $\operatorname{Re}(-ie^{i\mu}) = \operatorname{Re}(\sin\mu - i\cos\mu)$. Hence, if $f(z)$ is in $CV(i)$ and satisfies the condition (67), then

$$\textbf{(70)} \quad -ie^{i\mu}(1-2e^{-i\mu}\cos\nu z + e^{-2i\mu}z^2)f'(z) + i\cos\mu \ll \sin\mu\,\frac{1+z}{1-z} \ll \sin\mu\left(1+\sum_{n=1}^{\infty} 2z^n\right).$$

Finding sharp bounds for $|a_n|$ as a function of μ and ν seems to be an unpleasant task that is still to be finished.

Let $CVR(i)$ denote the subset of $CV(i)$ of functions for which all the coefficients are real. If $f(z)$ is in $CVR(i)$, then by the symmetry of $f(E)$, this domain is in $D_2(i)$ and hence Theorems 18 and 19 do not apply to $f(z)$. Further, if $f(z)$ is in $CVR(i)$, then max Re $(f(z))$ occurs at $z = 1$ and min Re $(f(z))$ occurs at $z = -1$. Thus, $f(z)$ has both the usual normalization and the one required by Hengartner and Schober. Consequently, by Theorem 15, we have

$$\text{(71)} \qquad \operatorname{Re}\,(1 - z^2)f'(z) \geq 0$$

for z in E. We can obtain the same condition from the inequality (67) by setting $\mu = \nu = \pi/2$.

Theorem 21. **PLE.** Robertson 1936 [1-B-1142]. If $f(z)$ is in $CVR(i)$, then there is a $\mu(\phi)$ in $M(0, \pi)$ such that

$$\text{(72)} \qquad f'(z) = \frac{1}{\pi}\int_0^{\pi} \frac{1}{1 - 2z\cos\phi + z^2}\, d\mu(\phi)$$

and

$$\text{(73)} \qquad f(z) = \frac{1}{2\pi i}\int_0^{\pi} \frac{1}{\sin\phi} \ln \frac{1 - ze^{-i\phi}}{1 - ze^{i\phi}}\, d\mu(\phi).$$

The representation (72) follows from Theorem 5, equation (20), and the condition (71).

Note that if $\mu(\phi)$ has a nonzero jump at 0 or π, then the integral in (73) is not defined. In this case we return to (72) and add a term $c/(1 \pm z)^2$ which on integration contributes $cz/(1 \pm z)$ to the integral in (73).

Theorem 22. **PLE.** Robertson 1936 [1-B-1144]. If $f(z) = z + a_2z^2 + \cdots + a_nz^n + \cdots$ is in $CVR(i)$, then $|a_n| \leq 1$ for all n,

$$\frac{1}{(1+r)^2} \leq |f'(z)| \leq \frac{1}{(1-r)^2},$$

and

$$\frac{r}{1+r} \leq |f(z)| \leq \frac{r}{1-r}.$$

All of these inequalities are sharp.

Typically-real functions and functions convex in the direction of the i-axis and having real coefficients are connected by

Theorem 23. Robertson 1936 [1-B-1144]. Set $f(z) = z + \cdots$. Then $f(z)$ is in $CVR(i)$ if and only if $zf'(z) \equiv g(z)$ is in TR.

Proof. This follows immediately from equations (21) and (72). ∎

5. FUNCTIONS THAT ARE STARLIKE IN ONE DIRECTION

We begin by looking for an Alexander type theorem for functions in the set $CV(i)$. It is a simple matter to define a new set $ST(r)$ of functions by imposing the condition that $f(z)$ is in $ST(r)$ if and only if there is some $g(z)$ in $CV(i)$ such that

$$f(z) = zg'(z). \tag{74}$$

With this definition an Alexander type theorem becomes trivally true. The question that must be answered is this: What geometric properties must $f(z)$ have if $f(z)$ is in $ST(r)$? As the symbol suggest, a function in $ST(r)$ is called starlike in the direction of the real axis, and this title suggests that the real axis meets $f(E)$ in either a single segment or a half-line or is contained in $f(E)$. For example, the function $z/(1-z^2)$ is univalent in E and the image of E contains the real axis.

To look closer at the source of the symbol $ST(r)$, let us examine a function $g(z)$ in $CV(i)$ that is also regular on the boundary. Then (see the condition (39)) there is an arc $\alpha \leq \theta \leq \beta$ of the unit circle such that $\operatorname{Im} zg'(z) = \operatorname{Im} f(z) \geq 0$ on that arc, and $\operatorname{Im} zg'(z) = \operatorname{Im} f(z) \leq 0$ on the complementary arc. This suggests that $f(z)$ maps E as indicated in Figure 7. In this picture the real axis meets $f(E)$ in a single line segment. By adding a suitable condition we can prove

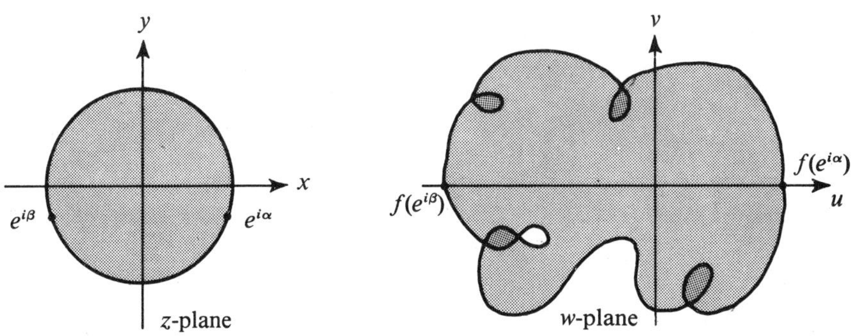

FIGURE 7.

Theorem 24. Suppose that $g(z)$ is in $CV(i)$ and is regular in $|z| \leq 1$. Then $f(z) \equiv zg'(z)$ has the property that $f(E)$ meets the real axis in a line segment I and covers I univalently.

This means that if u_0 is an interior point of I, then the equation $f(z) = u_0$ has exactly one root z_0 in E.

Proof. Since $g(z)$ is in $CV(i)$, we have

$$\frac{\partial}{\partial\theta} \operatorname{Re} g(z) = -\operatorname{Im} zg'(z) = -\operatorname{Im} f(z). \tag{75}$$

Hence the imaginary part of $f(z)$ changes sign exactly two times on $|z| = 1$. Thus $f(z)$ maps $|z| = 1$ onto a closed curve that meets the real axis in two points, as indicated in Figure 7, and hence cuts off some segment I of the real axis. If u_0 is any interior point of I, then $\Delta\arg\,(f(e^{i\theta}) - u_0) = 2\pi$ and hence $f(z) = u_0$ has exactly one solution in E. ∎

It is possible to weaken the condition of regularity on the boundary of E, but the generalization of Theorem 24 to the full set $CV(i)$ is not possible as we will see in Problem 71.

One might attempt to prove the full generalization by considering $f(\varrho z/\varrho)$ and taking the limit as $\varrho \to 1^-$. However just as in the set $CV(i)$ such an attempt must fail and for the same reason (see the next theorem). But first a little notation.

We let $\widetilde{ST}(L)$ denote the set of all functions $f(z) = z + \cdots$ that are regular in E and for which the intersection of $f(E)$ with the line L: $w = te^{i\alpha}$ is a connected set that is covered univalently by $f(E)$. If L is the real axis then we denote the set by $\widetilde{ST}(r)$. We will see in Problem 71 that $ST(r) \neq \widetilde{ST}(r)$. However the exact relation between the sets $ST(r)$ and $\widetilde{ST}(r)$ is an open problem that deserves further investigation.

Theorem 25. Hengartner and Schober [3-B-357]. There are functions $f(z)$ in $\widetilde{ST}(r)$ such that $f(\varrho z)/\varrho$ is not in $\widetilde{ST}(r)$ for some values of $\varrho < 1$.

Proof. Let $f(z)$ map E conformally onto the domain shown in Figure 8, with $f(0) = 0$. It is an easy matter to contract or expand the domain so that $f'(0) = 1$, and we consider this done. It is clear that $\operatorname{Im} f(e^{i\theta})$ changes sign exactly twice for θ in $(0, 2\pi]$. On the other hand there is some ϱ slightly less than 1 for which $f(C_\rho)$ will have the shape of the dashed curve and $\operatorname{Im} f(\varrho e^{i\theta})$ will change sign more than twice (in the figure it will change sign sixteen times) for θ in $(0, 2\pi]$. Hence $f(z)$ is in $\widetilde{ST}(r)$, but $f(\varrho z)/\varrho$ is not in $\widetilde{ST}(r)$ for some $\varrho < 1$.

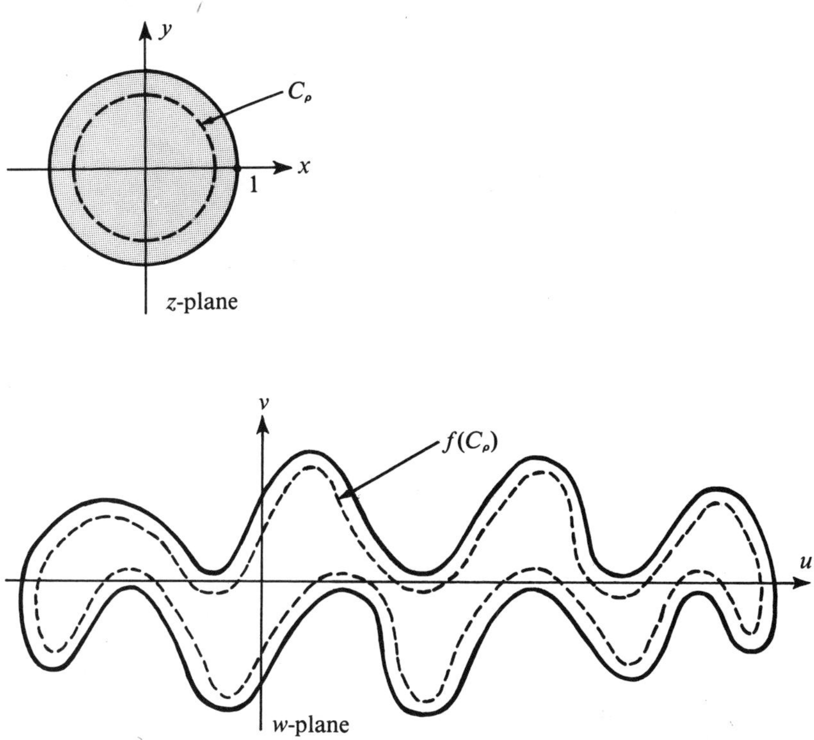

FIGURE 8

This surprising behavior of a harmonic function was noted much earlier by Bazilevič and Korickii 1953 [1-B-80], who gave a far more complicated and difficult proof in a different situation. Their objective was to prove that a curve might be "nearly" convex but that its level curves could be very "badly" not convex. Stated accurately, they proved

Theorem 26. PWO. Let $2N$ be a fixed even integer. There is a function $F(z)$ in S for which the boundary curve of $f(E)$ has exactly two inflection points, but one of the level curves, $F(C_\rho)$ with $\varrho < 1$, has at least $2N$ inflection points.

The concept of the proof is clear. We ask for a function $F(z)$ for which

$$\operatorname{Re}\left(1 + \frac{zF''(z)}{F'(z)}\right) = \operatorname{Im} f(z)$$

where $f(z)$ maps E onto a domain of the type indicated in Figure 8. Of course, there is a little labor involved in proving that $F(z)$ is univalent.

We return to the set $ST(r)$. The following four theorems are due to Robertson 1936 [1-B-1142].

Theorem 27. The function $f(z)$ is in $ST(r)$ if and only if there is a μ and ν in $[0, \pi]$ and a $p(z)$ in P such that

$$f(z) = z\,\frac{e^{-i\mu}(\cos\mu + i\sin\mu\, p(z))}{1 - 2\cos\nu e^{-i\mu}z + e^{-2i\mu}z^2}. \tag{76}$$

Proof. This follows from the definition of $ST(r)$, equation (74), and equation (67) in Theorem 20. ■

Theorem 28. PLE. If $f(z) = z + \Sigma_{n=2}^{\infty} a_n z^n$ is in $ST(r)$ then $|a_n| \leq n^2$ for all $n \geq 2$ with equality if and only if

$$f(z) = \frac{z + \eta z^2}{(1 - \eta z)^3}, \qquad \eta = \pm i. \tag{77}$$

We can combine Theorem 28 and equation (74) to obtain immediately

Theorem 29. If $g(z) = z + \Sigma_{n=2}^{\infty} b_n z^n$ is in $CV(i)$, then $|b_n| \leq n$ for all $n \geq 2$, with equality if and only if

$$g(z) = \frac{z}{(1 - \eta z)^2}, \qquad \eta = \pm i.$$

This theorem can also be proved directly from the inequality (67).

Theorem 30. PWO. If $f(z) = z + \Sigma_{n=1}^{\infty} a_{2n+1} z^{2n+1}$ is an odd function in $ST(r)$, then for $n \geq 1$ we have $|a_{2n+1}| \leq 2n + 1$, and this inequality is sharp for each n. The extremal function is

$$F(z) = \frac{z + z^3}{(1 - z^2)^2}. \tag{78}$$

Using the representation formula (76) for functions in $ST(r)$, Privalova and Saran 1964 [1-B-A-121] obtained sharp bounds in terms of μ and ν for (a) $\operatorname{Re} f(z)$, (b) $\operatorname{Im} f(z)$, and (c) $|f(z)|$ for $f(z)$ in $ST(r)$. The expressions for these bounds are extremely complicated.

6. FUNCTIONS WITH POSITIVE REAL PART AND REAL COEFFICIENTS

We recall from Theorem 4 that the transformations

$$\text{(79)}\qquad \frac{z}{1-z^2}p(z) = f(z) \qquad \text{and} \qquad p(z) = \frac{1-z^2}{z}f(z)$$

give a one-to-one correspondence between the set TR and PR. Now (as long as the coefficients are real) the sets STR, CVR, and $CVR(i)$ are subsets of TR and as such the transformation (79) carries each of these three sets into well-defined subsets of PR. One naturally asks for some description or properties of these three sets in PR (other than the definition). This difficult problem has been thoroughly investigated by Styer and Wright 1977 [3-B-895], and the reader is referred to this paper for a detailed statement of their results.

PROBLEMS

1. Derive the power series (10) by noting that

   (80)

   $$\tilde{k}_\phi(z) \equiv \frac{z}{1-2z\cos\phi+z^2} = \frac{1}{2i\sin\phi}\left[\frac{1}{1-ze^{i\phi}} - \frac{1}{1-ze^{-i\phi}}\right].$$

2. Prove that $\tilde{k}_\phi(z)$ is starlike in E by showing that $z\tilde{k}'_\phi(z)/\tilde{k}_\phi(z)$ is regular in E and the real part vanishes on the boundary of E. Is $\tilde{k}_\phi(z)$ in TR?
3. Show that $\tilde{k}_\phi(z)$ can be obtained from a Schwarz-Christoffel transformation. See equation (8.51).
4. Obtain $\tilde{k}_\phi(E)$ by examining the image of the boundary of E, using the form

   $$\tilde{k}_\phi(z) = \frac{1}{z+\dfrac{1}{z}-2\cos\phi}.$$

5. Use the process indicated in Problem 8.14 to show that $\tilde{k}_\phi(z)$ is a starlike function.
6. Prove that if $f(z)$ is in TR, then all the coefficients in (1) must be real. Prove that the converse is false.
7. Prove that the set TR is convex, namely, if $0 < t < 1$ and $f(z)$ and $g(z)$ are in TR, then $F(z) \equiv tf(z) + (1-t)g(z)$ is in TR.

8. Under the conditions of Problem 7 prove that the powers can be defined so that

$$G(z) = f^t(z)g^{1-t}(z) \tag{81}$$

is in TR. Rogosinski 1932 [1-B-1180] for Problems 8 through 19.

9. If $-1 < \lambda < 1$ and $f(z)$ is in TR, does the function

$$F(z) \equiv \left[f\left(\frac{z+\lambda}{1+\lambda z}\right) - f(\lambda)\right]\frac{1}{f'(\lambda)(1-\lambda^2)} \tag{82}$$

belong to TR?

10. If $f(z)$ is in TR and omits the real value γ, show that

(83)

$$F(z) \equiv \frac{\gamma f(z)}{\gamma - f(z)} = z + \left(a_2 + \frac{1}{\gamma}\right)z^2 + \left(a_3 + \frac{2a_2}{\gamma} + \frac{1}{\gamma^2}\right)z^3 + \cdots$$

is also in TR.

11. If $f(z)$ is in TR, show that $-f(-z)$ is in TR.

12. Suppose $f(z)$ is in TR and n is a positive integer. Is $\sqrt[n]{f(z^n)}$ an element of TR?

13. What can you say about $(\Pi_{k=1}^n f_k(z))^{1/n}$ if each $f_k(z)$ is in TR?

14. Prove that if $f(z)$ is in TR, then

$$\int_0^z \frac{f(\zeta)}{\zeta}\, d\zeta$$

is in TR.

15. What can you say about $\dfrac{2}{z}\displaystyle\int_0^z f(\zeta)\, d\zeta$?

16. Suppose that $f(z)$ is in $TR^{(2)}$ (an odd function in TR). Show that $f(z)$ has an integral representation

$$f(z) = \frac{1}{2\pi}\int_0^{2\pi} \frac{z+z^3}{1-2z^2\cos 2\phi + z^4}\, d\mu(\phi), \tag{84}$$

where $\mu(\phi)$ is in $M(0, 2\pi)$.

17. Use Problem 16 to prove that if

$$f(z) = z + \sum_{n=1}^{\infty} a_{2n+1}z^{2n+1}$$

is in $TR^{(2)}$, then $|a_{2n+1}| \leq 2n+1$ and this inequality is sharp for all n.

18. Suppose that $f(z)$ is in $SR^{(2)}$ (an odd function in SR). Prove that if z is in the kth quadrant (k = I, II, III, or IV), then $f(z)$ is in the kth quadrant.

19. Under the conditions of Problem 18 prove that $f(z)/z$ is in PR. Hence $|a_{2n+1}| \leq 2$ for each $n > 1$. This inequality is not sharp.

20. Recall from Problem 7.22 that

$$\frac{z}{\sqrt{1 - 2z^2 \cos \alpha + z^4}} = z + \sum_{n=1}^{\infty} P_n(\cos \alpha) z^{2n+1},$$

where P_n is the nth Legendre polynomial. Use this, Problem 2, inequality (26), and the remark that $\sqrt{f(z^2)}$ is in TR when $f(z)$ is in TR to prove that

$$1 + P_1(\cos \alpha) + P_2(\cos \alpha) + \cdots + P_n(\cos \alpha) \geq 0$$

for all real α. Fejér 1925 [1-B-292].

21. If $g_1(z)$ is an even function in PR, prove there is some function $g(z)$ in PR such that $g_1(z) = g(z^2)$. The converse is trivial. Rogosinski 1932 [1-B-1180] for Problems 21 through 24.

22. Prove that the mapping in Theorem 4 takes an odd function in TR into an even function in PR and conversely.

23. Prove that if $f_1(z)$ is in $TR^{(2)}$, then it has the form

$$f_1(z) = \frac{1 + z^2}{z} f(z^2)$$

for some $f(z)$ in TR. Conversely, if $f(z)$ is in TR, then $f_1(z)$ is in $TR^{(2)}$. Use this result to obtain the integral representation

$$\text{(85)} \quad f_1(z) = \frac{1}{2\pi} \int_0^{2\pi} \frac{z + z^3}{1 - 2z^2 \cos \phi + z^4} \, d\mu(\phi), \quad f_1(z) \in TR^{(2)}.$$

Compare this formula with the one obtained in Problem 16 where we have $\cos 2\phi$ in place of $\cos \phi$. Gel'fer 1964 [1-B-A-41] has obtained sharp bounds for $|f(z)|$, $|f'(z)|$, and $|zf'(z)/f(z)|$ for the set $TR^{(2)}$.

24. Suppose that $f(z)$ in TR has the power series (1). Prove that

$$|a_{n+2} - a_n| \leq 2, \qquad n \geq 1,$$

$$1 + a_2 + a_3 + \cdots + a_{n-1} + \frac{a_n}{2} \geq 0, \qquad n \geq 1,$$

$$n + (n-1)a_2 + \cdots + 2a_{n-1} + a_n \geq 0, \qquad n \geq 2.$$

Hint: Use Theorems 4 and 6 together with equations (1.6) and (1.12).

25. In Theorem 7 find $f_1(z)$ when $f(z)$ is (a) $k(z)$, (b) $z/(1-z)$, and (c) z. Is $f_1(z)$ always univalent?

★ **26.** If $f(z)$ is in TR, prove that

$$|f'(z)| \le \begin{cases} \left|\dfrac{1+z}{(1-z)^3}\right|, & \text{if } \operatorname{Re}\left(\dfrac{1+z^2}{z}\right) \ge 2, \\ \left|\dfrac{1-z^2}{z^2}\right|\left|\operatorname{Im}\dfrac{1+z^2}{z}\right|^{-2}, & \text{if } \left|\operatorname{Re}\left(\dfrac{1+z^2}{z}\right)\right| \le 2, \\ \left|\dfrac{1-z}{(1+z)^3}\right|, & \text{if } \operatorname{Re}\left(\dfrac{1+z^2}{z}\right) \le -2. \end{cases} \tag{86}$$

Goluzin 1950 [1-B-413]. See Gel'fer [1-B-A-41] for a similar bound for $|Q_{ST}(f)|$.

27. Derive the following integral representations for the set of functions indicated.

$$\ln\sqrt{\frac{1}{f'(z)}} = \frac{1}{2\pi}\int_0^{2\pi} \ln(1 - ze^{-i\phi})\, d\mu(\phi), \qquad f(z) \in CV. \tag{87}$$

$$f^{(n)}(z) = \frac{1}{2\pi}\int_0^{2\pi} \frac{2(n!)e^{-in\phi}}{(1-ze^{-i\phi})^{n+1}}\, d\mu(\phi), \qquad f(z) \in P. \tag{88}$$

★ **28.** Let $TR(E-0)$ denote the set of functions

$$F(z) = -\frac{1}{z} + b_0 + \sum_{n=1}^{\infty} b_n z^n \tag{89}$$

that are regular, are not zero, and are typically-real in $E-0$. By definition $F(z)$ satisfies the condition (11). Prove that functions in this set have the integral representation

$$F(z) = \frac{1}{2\pi}\int_0^{2\pi} \frac{-(1-z^2)^2}{z(1-2z\cos\phi + z^2)}\, d\mu(\phi).$$

Thus Theorem 8 can be applied to this set of functions. Gel'fer 1954 [1-B-372] used this representation to obtain sharp bounds for $|b_n|$. Remizova 1963 [1-B-1125] analyzed the region of values for $F(z)$ in $TR(E-0)$ and found sharp bounds for $|F(z)|$, Im $F(z)$, Re $F(z)$, and $|F'(z)|$. See Goodman 1956 [1-B-436] and the last part of Chapter 17.3.

29. Let $TR((-\infty, 0])$ denote the set of functions

(90) $$F(z) = (z-1) + \sum_{n=2}^{\infty} b_n(z-1)^n$$

that are typically-real in the entire complex plane minus the cut along the negative real axis from $-\infty$ to 0. By a transformation of equation (21) prove that $F(z)$ is in this set if and only if

(91) $$F(z) = \int_0^1 \frac{z-1}{1+t(z-1)}\, d\mu(t)$$

where $\mu(t)$ is in $M(0, 1)$. Merkes 1959 [1-B-913] for Problems 29 through 32.

30. Find maximum and minimum values for b_n for the set of functions defined in Problem 29.

★ **31.** Show that each function in the class defined in Problem 29 is univalent in the right half-plane, $|\arg z| < \pi/2$.

32. Prove that the domain of univalence found in Problem 31 is sharp.

33. Use Problems 31 and 32 to find the largest domain D in E such that every function in TR is univalent in D. This domain is called the *domain of univalence* for the set TR. Goluzin 1950 [1-B-413, p. 212].

34. Prove that for each point z_1 on the boundary of the domain found in Problem 33 there is an $f(z)$ in TR for which $f'(z_1) = 0$.

35. Let $U(z) = \operatorname{Re} Q_{ST}(z)$ and suppose that $f(z) = z + \Sigma_2^\infty a_n z^n$ is regular in E. Prove that if there is a constant K such that $U(z) > K$ or $U(z) < K$ for z in E, then $f(z)$ has only one zero in E, the one at the origin.

★ **36.** Let $\Gamma_r = f(C_r)$ be a curve that does not pass through the origin. Let $n(\Psi)$ be the number of times that a line (not a ray) through the origin, that makes an angle Ψ with the positive real axis, meets the curve Γ_r. Recall that

(92) $$U(z) \equiv \operatorname{Re} \frac{zf'(z)}{f(z)} = \frac{d}{d\theta} \arg f(re^{i\theta}).$$

Prove that

(93) $$\int_0^{2\pi} |U(re^{i\theta})|\, d\theta = \int_0^{\pi} n(\Psi)\, d\Psi.$$

Suppose that $f(z)/z = 1 + \cdots$ is regular and not zero in E. Prove that if

(94) $$\int_0^{2\pi} |U(re^{i\theta})|\, d\theta < 4\pi$$

for every r in (0, 1), then $f(z)$ is starlike in some direction (not necessarily in the direction of the real axis). Note that $|a_n| \leq n^2$ is still true for a function that satisfies these conditions. Draw a figure that will indicate that 4π is the best possible constant in (94).

★ **37.** Suppose that $f(z) = z + \cdots$ is regular in E. Using the notation and results of Problems 35 and 36, prove the following theorem. If there is a constant $\alpha > 3/2$ such that in E

$$\text{(95)} \qquad -\frac{\alpha}{2\alpha - 3} < \operatorname{Re} \frac{zf'(z)}{f(z)} < \alpha,$$

then $f(z)$ is starlike in some direction. Umezawa 1952 [1-B-1428]. Hint: Let $C^{(1)}$ be that part of C_r on which $U(z)$ is positive and let $C^{(2)}$ be the complement. Let x_1 be the length of $C^{(1)}$ and set

$$y_1 \equiv \int_{C^{(1)}} U(re^{i\theta})\, d\theta, \qquad -y_2 \equiv \int_{C^{(2)}} U(re^{i\theta})\, d\theta.$$

Prove that if $y_1 \geq 3\pi$, we obtain the contradiction $y_2 \geq \pi$ and $y_2 < \pi$. Hence $y_1 < 3\pi$ and we obtain (94).

★ **38.** Use the result of Problem 37 to obtain the following new condition that is sufficient for a function to be univalent. If $f(z) = z + \cdots$ is regular in E and there is a constant $\alpha \geq 3/2$ such that in E

$$\text{(96)} \qquad -\frac{\alpha}{2\alpha - 3} < \operatorname{Re} \left(1 + \frac{zf''(z)}{f'(z)}\right) < \alpha,$$

then $f(z)$ is convex in some direction and is univalent in E. The special values $\alpha = \infty$, 3, 2, and 3/2 give four different conditions sufficient for the univalence of a function. These special cases were first proved by Ozaki 1941 [1-B-1022]. The condition (96) is due to Umezawa [1-B-1428]. It has been generalized by Ogawa 1961 [1-B-1007, p. 11] and investigated in more detail by G. M. Shah 1973 [2-B-1301], who obtained bounds for $|a_n|$ and $|f(z)|$ as a function of α.

39. Prove that if $f(z)$ is in TR, then

$$\left| \frac{zf'(z)}{f(z)} \right| \leq \frac{1+r}{1-r}$$

and the result is sharp. Gel′fer 1964 [1-B-A-41].

40. Prove that the function $F(z)$ defined by equation (41) has the mapping properties claimed for it. See Problem 2.32.

41. Check that equation (43) is correct.

42. Show that if $r = 0.55$ and $\alpha = 30°$, then $u'(\theta)$ has four changes of sign in $(0, 2\pi)$.

43. Prove that if $f(z)$ is in $\widetilde{CV}_2(i)$ and $\text{Re}\,(1 - z^2)f'(z) = 0$ for just one point in E^+, then $f(z)$ must have the form given in equation (54).

44. Prove Theorem 16.

45. Prove that in E, $\text{Re}\,(1 - z^2)f'(z) \geq 0$ for the function defined by equation (60).

46. Prove that $f(z) = iz/(1 - z)^2$ satisfies the inequality (65).

47. Let $G(z) = F(\bar{B}z)$ where $F(z)$ is defined by equation (41). Prove that in E we have $\text{Im}\,(1 - z)^2 G'(z) \leq 0$ as Theorem 19 predicts.

48. Find μ and ν in Robertson's Theorem, inequality (67), for $f_1(z)$ and $f_2(z)$ defined by equation (48).

49. Prove Theorems 21 and 22.

50. Prove that if $f(z)$ is in $CV(i)$ and $\mu = 0$ in (67), then

$$f(z) = \frac{i}{2\sin\nu} \ln \frac{1 - e^{i\nu}z}{1 - e^{-i\nu}z}, \qquad \nu \neq 0, \pi. \tag{97}$$

See Problem 62.

51. Prove Theorems 28 and 29, namely, $|a_n| \leq n^2$ for $f(z)$ in $ST(r)$, and $|b_n| \leq n$ for $g(z)$ in $CV(i)$.

52. Let $TR(i)$ denote the subset of TR of functions $f(z)$ for which $f(iz)/i$ is also in TR. In other words $f(z)$ is also typically-real with respect to the i-axis so that for z in E we have $(\text{Re}\, f(z))(\text{Re}\, z) \geq 0$.

If $f(z)$ is in $TR(i)$, prove that $f(z)$ is an odd function. Further, prove that $f(z)$ is in $TR(i)$ if and only if there is a $g(z)$ in TR such that $f(z) = (g(z^2))^{1/2}$. This new set $TR(i)$ was introduced in 1935 by Robertson [1-B-1140]. See also Goluzin 1950 [1-B-413, p. 216].

53. Prove that $SR^{(2)} \subset TR(i)$.

54. Suppose that $f(z)$ is in $TR^{(2)}$. Does this imply that $f(z)$ is in $TR(i)$?

55. Suppose that $f(z) = z + a_3 z^3 + a_5 z^5 + \cdots$ is in $TR(i)$. Prove that $-1 \leq a_3 \leq 1$. Are these bounds sharp? Robertson 1935 [1-B-1140] for Problems 55 and 56.

56. Under the conditions of Problem 55 prove that $|a_{2k-1}| + |a_{2k+1}| \leq 2$. Compare this result with the bound in Problem 17 for $TR^{(2)}$. This inequality was proved for the subset $SR^{(2)}$ by Dieudonné 1931 [1-B-242, p. 318] and in a simpler way by Takahashi 1933 [1-B-1368].

★ **57.** Under the conditions of Problem 56 prove that $1 \leq a_5 \leq 3/2$.

Goluzin 1950 [1-B-413, pp. 216–217]. Is this upper bound sharp? For further bounds in the set $TR(i)$ see Gel'fer 1964 [1-B-A-41].

Problems 58 through 61 give a certain extension of the idea of a typically-real function and lead to a proof that $|a_n| \leq n$ for a set of univalent functions not included in any one of the sets ST, TR, or $CV(L)$. It represents the work of De Bruijn 1941 [1-B-235] and Ozaki 1941 [1-B-1022, p. 79] who obtained the same results independently.

58. Let $f(z) = z + a_2z^2 + \cdots$ be regular in E and continuous in $\bar{E}$. Let w_0 be a fixed point in $f(E)$ such that $w_0 \neq f(e^{i\theta})$ for any real θ. Prove that there is a real number α such that $f(e^{i\alpha})$, w_0, and $f(-e^{i\alpha})$ lie on a line. Ozaki calls the line through these three points a *diametral line*. Hint: Consider $\Phi(\theta) \equiv \arg[(f(-e^{i\theta}) - w_0)/(f(e^{i\theta}) - w_0)]$ as a continuous function as θ runs from 0 to π.

The concept of a diametral line has been refined by Umezawa 1952 [1-B-1430] to a consideration of the equation

$$\arg(f(-e^{i\theta}) - w_0) = \arg(f(e^{i\theta}) - w_0) + p\pi.$$

59. We would like to assert in Problem 58 that the points $f(e^{i\alpha})$, w_0, and $f(-e^{i\alpha})$ lie on the diametral line in that order. Give an example to show that if $f(z)$ is not univalent, then such an assertion may be false. Prove that if $f(z)$ is univalent, there is an α for which $\Phi(\alpha) = \pi$.

60. Let $f(z)$ satisfy the conditions of Problem 58 and assume that there is a diametral line such that $f(e^{i\theta})$ is on "one side" of the line if and only if $\alpha < \theta < \alpha + \pi$. Prove that $|a_n| \leq n$ for all n. Here w_0 may lie outside $f(E)$. Note that a special case of this is the theorem for bounded typically-real functions. Hint: By a rotation set $g(z) = e^{-i\alpha}f(e^{i\alpha}z)$. Then there are real numbers μ and b such that on $|z| = 1$ we have $\operatorname{Im}\{g(z)(e^{-i\mu} - ib)\} \geq 0$ for $0 \leq \theta \leq \pi$ and $\operatorname{Im}\{g(z)(e^{-i\mu} - ib) \leq 0$ for $\pi \leq \theta \leq 2\pi$. Consequently $\operatorname{Re}\{-(z - 1/z)g(z)(e^{-i\mu} - ib)\} \geq 0$, first on $|z| = 1$, and then throughout $\bar{E}$. Then use the fact that $|a_2| \leq 2$ and the coefficient theorem for functions with positive real part.

61. Combine the results of Problems 59 and 60 to prove that if $f(z)$ is a bounded univalent function that is starlike with respect to some w_0 (either inside or outside $f(E)$), then $|a_n| \leq n$.

★ **62.** For $0 < \phi < \pi$, set

$$f(z) = \frac{1}{2i \sin\phi} \ln \frac{1 - ze^{-i\phi}}{1 - ze^{i\phi}}.$$

Prove that $\operatorname{Re} f(z)/z > 1/2$ in E. See Marx 1932 [1-B-909, pp. 59–61]. For a simple proof see Komatu 1961 [1-B-723].

63. Use the result in Problem 62 together with Theorem 21 to prove that if $f(z)$ is in $CVR(i)$, then $\operatorname{Re} f(z)/z > 1/2$ in E. Hence $f(z)/z \prec 1/(1-z)$.

64. Use Theorem 8.12 to prove that if $f(z)$ is in $CVR(i)$, then there is a $\mu(\phi)$ in $M(0, 2\pi)$ such that

$$f(z) = \int_0^{2\pi} \frac{z\, d\mu(\phi)}{1 - ze^{i\phi}}.$$

65. Following Mocanu's concept of an alpha-convex function, Krystyna Skalska 1980 [3-B-866] introduced the set of alpha-typically-real functions. A function $f(z) = z + \Sigma_2^\infty a_n z^n$, regular in E, is said to be alpha-typically-real if all the coefficients are real and $\operatorname{Re} J(\alpha, f) > 0$ in E, where

$$J(\alpha, f) \equiv (1-\alpha)\frac{1-z^2}{z} f(z) + \alpha(1-z^2) f'(z). \tag{98}$$

Theorem 4 states the if $f(z)$ is in TR, then $(1-z^2)f(z)/z$ is in PR and conversely. Further, Theorem 23 implies that if $f(z)$ is in $CVR(i)$, then $(1-z^2)f'(z)$ is in PR and conversely. This explains the selection of functions in (98). We let $\alpha - TR$ denote the set of all alpha-typically-real functions. For special values of α we have $0 - TR = TR$ and $1 - TR = CVR(i)$. Prove that if $f(z)$ is in $\alpha - TR$ and $\alpha \neq 0$, then there is a $p(z)$ in PR and an $F(z)$ in TR such that

$$f(z) = \frac{1}{\alpha} z^{1-(1/\alpha)} \int_0^z \zeta^{-1+(1/\alpha)} \frac{p(\zeta)}{1-\zeta^2}\, d\zeta,$$

or

$$f(z) = \frac{1}{\alpha} z^{1-(1/\alpha)} \int_0^z \zeta^{-2+(1/\alpha)} F(\zeta)\, d\zeta.$$

For more properties of the set $\alpha - TR$ see Jakubowski 1981 [3-B-389].

The next four problems are from the paper by Skalska.

66. Prove that if $0 \le \alpha \le \beta$ then $\beta - TR \subset \alpha - TR$. Hence for every $\beta \ge 0$ a function in $\beta - TR$ is also typically-real. Further, for every $\beta \ge 1$, we have $\beta - TR \subset CVR(i) \subset S$.

67. Find a function in TR that is not in $\alpha - TR$ for any $\alpha > 0$.

68. Find a function in $CVR(i)$ that is not in $\alpha - TR$ for any $\alpha > 1$.

69. Find a good upper bound for $|a_n|$ if $f(z)$ is in $\alpha - TR$ and $\alpha \geq 0$.

70. Prove Theorem 17. Hengartner and Schober 1970 [2-B-571]. If $|f'(0)| = 1$, then $f(E)$ covers the disk $|w - f(0)| < \ln \sqrt{2}$.

★ **71.** Prove that $ST(r) \neq \widetilde{ST}(r)$ by filling in the details in the following outline. Let $F(z)$ be defined by equation (41) and set $G(z) = zF'(z)$. Now $F(z)$ is in $CV(i)$ and hence $G(z)$ is in $ST(r)$. But $G(E) \cap \{\text{the real axis}\}$ is not a connected set if $0 < \alpha < \pi/6$. To see this, note that $G(e^{i\theta})$ is real and that $G'(z)$ has two zeros on $|z| = 1$. Hence $g(C_1)$ covers some interval on the real axis three times and hence $G(E)$ omits this interval. Then $G(z)$ is not in $\widetilde{ST}(r)$.

REFERENCES

REFERENCES NOT CONTAINED IN BERNARDI'S BIBLIOGRAPHY

1. O. P. Ahuja, *Certain generalizations of the Robertson function,* Yokohama Math. Jour., to appear.
1. S. D. Bernardi, *Bibliography of Schlicht Functions*, 1983, Mariner Publishing Co., 10927 North Dale Mabry, Tampa, FL. 33618.
1. J. G. Clunie and F. Holland, Preprint 1981.
1. Nelson Dunford and J. T. Schwartz, *Linear Operators*, 1958, Wiley Interscience Press, New York—London.
1. A. Gutzmer, *Ein Satz über Potenzreihen*, Math. Ann. vol. 32 (1888) pp. 596–600.
1. E. Jacobsthal, *Sur l'inversion d'une série de puissance I*, Kongelige Norske Videnskabers Selskab Forh. vol. 21 (1948) pp. 14–17.
1. F. Kamber, *Formulas exprimant les valeurs des coefficients des séries de puissances inverses*, Acta Math, vol. 78 (1946) pp. 193–204.

1. A. I. Markushevich, *Theory of functions of a complex variable*, English translation by Richard A. Silverman, 1965, Prentice-Hall, Englewood Cliffs, New Jersey.
1. M. L. Mogra and O. P. Ahuja, *On spiral-like functions of order α and type β*, Yokohama Math. Jour. vol. 29 (1981) pp. 145–156.
1. R. Nevanlinna, *Analytic Functions*, Springer Verlag, Berlin 1970.
1. V. C. Rogozin, *Sufficient conditions for a univalent solution of an inverse boundary value problem*, (Russian) Priklad. Mat. and Mech. vol. 22 (1958) pp. 804–807.
1. A. Sathaye and S. M. Shah, *The second coefficient of a function with all derivatives univalent*, Inter. Jour. of Math. and Math. Sciences, vol. 5 (1982) pp. 281–288.
1. Pavel G. Todorov, *Explicit formulas for the coefficients of α-convex functions, $\alpha \geq 0$.* Canadian Jour. Math. To appear.

ANSWERS

CHAPTER 2

2. $A = \csc(\pi/k)$. **3.** The strip $|\operatorname{Im} w| < \pi/4$.

6. $|a_2| \le 1/2$, $|b_2| \le |b_1|/2$.

7. No. Consider the Koebe function given in Problem 3.

14. $|a_n| \le (\sin \pi/k)^{n-1}(k-1)!/n!(k-n)! < \pi^{n-1}/n!$, for $k \ge n \ge 2$.

15. $|b_n| \le n|b_1|$.

16. $\phi(\zeta) = \zeta - \frac{1}{k}a_2 \frac{1}{\zeta^{k-1}} + \left(\frac{k+1}{2k^2} a_2^2 - \frac{1}{k}a_3\right)\frac{1}{\zeta^{2k-1}} + \cdots$.

17. If $\phi(\zeta)$ is in Σ, $f(z)$ will be univalent in E, but may have a pole in E.

18. $k_p(E)$ is the entire plane minus the interval $[-p/(1-p)^2, -p/(1+p)^2]$. $k_1(z) = z/(1-z)^2$.

19. $a_n = \Sigma_{k=0}^{n-1} p^{n-2k-1} \equiv A_n(p)$.

20. If $f(z)$ is in $S(p)$, then $|a_n| \le A_n(p)$. For some results related to this conjecture see Theorems 4.6 and 4.7. For more results about $S(p)$ see Chapter 17.4.

21. $k(c, E)$ is a sector symmetric with respect to the real axis. The vertex is at $-1/2c$ and the angle of the sector is $c\pi$.
22. The complex plane minus two slits along the imaginary axis from $\pm i/2$ to ∞. Apply the transformation (10) to the Koebe function with $k = 2$.
23. $\Sigma_{n=0}^{\infty} z^{2n+1}$. Conjecture: $|a_{2n+1}| \leq 1$. When $n = 1$, this conjecture is true (see Corollary 4.2). For every $n > 1$ this conjecture is false. Fekete and Szegö 1933 [1-B-304] proved that the true bound for $|a_5|$ is $1/2 + (1/e)^{2/3} \approx 1.0134$.
24. 1/2. **25.** 1/2. **26.** 1/3.
27. Yes. If $r = 1/k$, then $\sum_{j=2}^{k+1} \binom{k+1}{j} (j-1)r^j = 1$.
28. 1. **29.** $a(n+1) + bn < 2an + bn$. **30.** π
31. $h(e^{i\theta}) = e^{i\alpha} \sin(\alpha + \theta/2)/\sin(\theta/2)$.
32. $f(e^{i\theta}) = (-1 + e^{i(\alpha+\theta)} \cos\alpha)/2(1 - \cos\theta)$, $\alpha + \pi/2$. The slit end point is $w_0 = -(1 + i \cot\alpha)/4$.
33. $|a_n| \leq n$ with equality if and only if $c = 0$.
34. $F_n(z)$ is univalent in E and $F_n(E)$ is the entire complex plane minus the negative real axis and cuts along n (or fewer) arcs that terminate on the negative real axis.

CHAPTER 3

1. Yes.
3. Since $-2(1+\epsilon)^2 - 1 + 1 + 2(1-\epsilon)^2 < 0$, then (by Theorem 3) Γ cannot be a simple closed positive curve.
4. Suppose that the series (7) converges on C_r and $\Gamma = f(C_r)$. If $\Sigma_{-\infty}^{\infty} n|a_n|^2 < 0$, then either Γ has self-intersection points or Γ has a negative direction.
5. $a = R + 1/R$, $b = R - 1/R$. The line segment $-2 \leq u \leq 2$.
6. Area $= \pi(R^2 - 1/R^2) = \pi ab$. **9.** $A \leq 3\pi/2$.
12. See Theorem 12.1 **16.** $L(\alpha, r, f) = \int_0^r |f'(te^{i\alpha})|\, dt$.

CHAPTER 4

1. The complex plane minus the two slits along the real axis: $u \geq 1/(2-b)$ and $u \leq -1/(2+b)$.
2. $e^{\pm i\phi}$, where $\cos\phi = b/2$. **3.** $a_2 = b$, $a_3 = b^2 - 1$.

8. $|b_3| = |2a_2^2 - a_3| = |2a_2^2 - 2a_3 + a_3| \le 2|a_2^2 - a_3| + |a_3| \le 2+3.$
9. 3. **11.** 1, see Problem 7. **12.** $F(z) = z/(1-z^k)^{2/k}$.
13. Equality holds for $k(z, M)$.
15. If $M = 1$, then $k(z, M) \equiv z$ and $a_n = 0$ for $n > 1$.
17. If $f_n(z) \to f(z) \not\equiv 0$ uniformly in D and z_0 has the property that for each ϵ in $(0, \epsilon_0)$ there is an $N(\epsilon)$ such that in $|z - z_0| < \epsilon$ every function $f_n(z)$ with $n > N(\epsilon)$ has k zeros, then $f(z)$ has a kth order zero at z_0.
25. Consider the function $f(z) = k(z) + b_2 k^2(z)$, where $k(z)$ is the Koebe function.

CHAPTER 5

5. $h(z) = 1 - z$. **6.** See equation (4.16).
7. $z = w + w^6 + \sum_{k=2}^{\infty} \frac{5k(5k-1)(5k-2)\cdots(4k+2)}{k!} w^{5k+1}$. Set $w = 1/3$ to find a root of $z^5 - 3z + 1 = 0$.
8. Use $(\Sigma_0^\infty b_n z^n)(\Sigma_0^\infty a_n z^n) \equiv 1$. The Cauchy product will yield a set of linear equations relating the a's and the b's.

CHAPTER 6

2. $r = 1/2$. Put $a_2 = 0$ in Problem 1.
3. What value does $g(z) = \gamma f(z)/(\gamma - f(z))$ omit?
5. Suppose that the curve L is parametrized by $\varrho = \varrho(t)$ and $\theta = \theta(t)$, where $\zeta = \varrho e^{i\theta}$. Then

$$d\zeta = \frac{d\zeta}{dt}\,dt = (\varrho'(t)e^{i\theta} + i\varrho(t)e^{i\theta}\theta'(t))\,dt = e^{i\theta}(\varrho'(t) + i\varrho(t)\theta'(t))\,dt.$$

Hence $|d\zeta| \ge |\varrho'(t)|\,dt \ge d\varrho$, when we restrict the integration to those portions of L on which ϱ is increasing.
7. $|\arg k'(z)| \le |\arg(1+z)| + 3|\arg(1-z)|$.
8. $\arg(F(z)/z)$ is unbounded in E.
10. $r/(1+r^k)^{2/k} \le |f(z)| \le r/(1-r^k)^{2/k}$.
11. $(1-r^k)/(1+r^k) \le |Q(z)| \le (1+r^k)/(1-r^k)$.
12. $(1-r^k)/(1+r^k)^{1+2/k} \le |f'(z)| \le (1+r^k)/(1-r^k)^{1+2/k}$.
13. Suppose that $a_2 \ge 0$ and a_2 is fixed. Then each point of the circle $\gamma = 1/(2e^{i\beta} - a_2)$ is achieved. For fixed β and a_2, the function $f(z) = z/(1 - a_2 z + z^2 e^{2i\beta})$ has the required second coefficient, and $f(e^{-i\beta}) = \gamma$, the required omitted value.
14. Use Theorems 3 and 6.

16. For each fixed r in $(0, 1)$ the extremal function is $-[\ln(1-rz)]/r$.
17. The extremal function is $z/(1-rz)$.
23. Map the exterior of the sector onto E and use Schwarz's Lemma.
24. See Theorem 12.19. **25.** $a_2 = \pi/2$.
26. After normalization $c(f^{(k)}(z) - f^{(k)}(0))$ is in ES.
27. $|a_k| \leq (2A_2)^{k-1}/k!$, $|f(z)| \leq (e^{2A_2 r} - 1)/2A_2$ for all z.

CHAPTER 7

7. Yes. Any rotation of $f(z) = -z - 2\ln(1-z)$.
8. $f'(E_r)$ must lie in the disk described in Theorem 7.
9. $-r + 2\ln(1+r) \leq |f(z)| \leq -r - 2\ln(1-r)$.
11. $|a_n| \leq 2(a-\alpha)$ for $n = 1, 2, 3, \ldots$.
$f(z) = \alpha + bi + (a-\alpha)(1+z)/(1-z)$.
12. Center at $bi + \alpha + (a-\alpha)(1+r^2)/(1-r^2)$, radius is $2(a-\alpha)r/(1-r^2)$.
13. $f'(z) \prec (1+z)^\beta/(1-z)^\beta$.
14. $|f(z)| \leq \int_0^r (1+t)^\beta/(1-t)^\beta\, dt < 2^\beta/(1-\beta)$. For some generalizations see Biernacki 1934 [3-B-114]. If $\beta \geq 1$, then $|f(z)|$ may be unbounded in E.
15. Integrate $[\operatorname{Re} f(z)]^2$ on a circle C_r.
$F(z) = \dfrac{\delta i}{\pi}\ln\dfrac{1+z}{1-z} = \dfrac{2\delta i}{\pi}\sum_1^\infty \dfrac{z^n}{2n-1}$ satisfies the hypotheses, and

$$\sum_1^\infty \frac{4\delta^2}{\pi^2}\frac{1}{(2n-1)^2} = \frac{4\delta^2}{\pi^2}\cdot\frac{\pi^2}{8} = \frac{\delta^2}{2}.$$

23. Consider $1/g^2(z)$. The half-plane $\operatorname{Re} w > 0$ minus the slit from ∞ to $|1/\cos(\alpha/2)|$.
29. Set $c = 1$ in equation (71).
31. If $p(z)$ is in P and $p(z) = 1 + \Sigma_1^\infty p_n z^n$, then for
$n \geq 2$, $|p_{n-1} - 2p_n + p_{n+1}| \leq 2\operatorname{Re}(2-p_1) \leq 2|2-p_1|$ and
$|2 - 2p_1 + p_2| \leq 2\operatorname{Re}(2-p_1) \leq 2|2-p_1|$.
33. (a) $q(z) \equiv 0$. (b) Yes, $p(z) = (1-z^2)/(1-z+z^2)$, and this $p(z)$ is in P, set $\cos\alpha = 1/2$ in Problem 20.
37. See Problem 9.3. $|p_n| \leq 2(1-\alpha)$, $P(z) = [1+(1-2\alpha)z]/(1-z)$. McCarty 1972 [2-B-883] has studied the subset of $P(\alpha)$ in which $|p_1|$ is fixed.
38. $p(z) = (Mb(z) + M)/(M - (M-1)b(z))$, $b(z)$ in B_0.
39. No. It is a bad conjecture. If $M = 1$, then $m = 0$ and $P(z) = 1 + z$,

so that 0 is the conjectured upper bound for $n > 1$ in $\widetilde{P}(1)$. Clearly, this is wrong. Goel 1967 [2-B-440] proved a theorem that is equivalent to the correct bound, $|p_n| \leq 2 - 1/M$ where $(1 + z^n)/(1 - mz^n)$ is an extremal function. See Goel 1970 [2-B-446] for some minor corrections to [2-B-440]. See also Janowski 1970 [2-B-642, p. 176]. If we assume that in addition $f(z)$ is univalent in E, then the determination of the sharp upper bound for $|p_n|$ seems to be a difficult problem. A proof that $|p_n| \leq 2 - 1/M$ can be given using Clunie's method; See Problem 17.33.
46. In Theorem 22, replace z by z^k.
47. The open disk with diameter end points $w_1 = (1 - r^k)/(1 + r^k)$, and $w_2 = 1/w_1$.

CHAPTER 8

1. $1 + zf''(z)/f'(z) = (d - cz)/(d + cz)$. Re $[(d - cz)/(d + cz)] > 0$ in E if and only if $|d/c| \geq 1$.
2. The inequality (7) is not true in E because (7) has a pole where $k(z_0) = w_0$. But for the reciprocal

$$\text{Re}\left(\frac{f(z) - w_0}{zf'(z)}\right) = \text{Re}\left[\left(\frac{1 - z}{1 + z}\right)\left(1 + 2w_0 - w_0(z + z^{-1})\right)\right] > 0$$

in a ring domain $R < |z| < 1$, for a suitable R.
7.

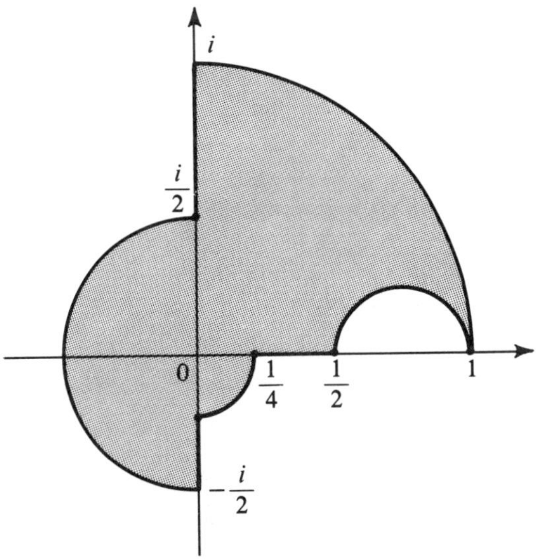

12. If and only if $f(z)$ is a rotation of $k(z)$.
13. If $f(z) \in CV$, then $|\arg f'(z)| \leq 2 \arcsin r$. For a proof, apply Theorem 7.7 to $g(z) \equiv 2\sqrt{f(z)} - 1 \in P$. Equality occurs in the bound iff $f(z)$ is a rotation of $z/(1-z)$.
14. Consider $z(f(z)/z)^{\lambda}(g(z)/z)^{\mu}$. If $f(z) = z/(1-z)$ and $g(z) = z/(1+z)$, $\lambda = \mu = 1/2$, then $F(z) = z/\sqrt{1-z^2}$ and $F(z) \notin CV$.
15. The domain bounded by the hyperbola $2v^2 - 2u^2 = 1$.
16. Show that $\arg F(z)$ is constant on each arc $e^{i\theta}$, where $\phi_k < \theta < \phi_{k+1}$. $F(E)$ is the entire plane minus m radial slits. The lengths and locations of the slits depend on the parameters $\phi_1, \phi_2, \ldots, \phi_m$, but the angle between slits is $\mu_k \pi$. The Koebe function is the special case $m = 1, \mu_1 = 2$, $\phi_1 = 0$.
17. Both f and g are in all three sets.
18. $\sqrt{2} - 1$. One extremal function is $f(z) = -z - 2\ln(1-z) \in P'$.
19. Using the result in Problem 7.36, prove that $\mathrm{Re}\,(zf'(z)/f(z)) \geq (1 - r^2 - 2r)/(1 - r^2)$. Then $f(z)$ is starlike in $|z| \leq R_{ST} = \sqrt{2} - 1$, hence, univalent in the same disk. An extremal function is $f(z) = (z + z^2)/(1-z)$ for which $f'(-R_{ST}) = 0$.
22. Set $f(z) = z$, $\zeta_0 = z_0$ in Problem 21.
23. Let $g(z) = z^n$, D: $0 > \arg z > -\pi/2n$, $n \geq 1$.
26. $g(z) = z^n$, $n \geq 1$, D: $\pi/2n < \arg z < \pi/n$.
28. Yes. Set $w_0 = 0$ for the starlikeness. For Γ_1 and Γ_2 set $F(z) = cf'(z)/g'(z)$, $c \neq 0$. Then $F(z)$ is starlike with respect to the origin on Γ_k if and only if $f(z)$ is convex on Γ_k. For Γ_3 and Γ_4 set $F(z) = cf'(z)g(z)/g'(z)$ with the same conclusion.
29. $f(z) = a_0 + a_k z^k$, $a_k \neq 0$, $k = 1, 2, 3, \ldots$. **30.** Yes.
31. No. See Problem 15.
32. $(1 + r^k)^{-2/k} \leq |f'(z)| \leq (1 - r^k)^{-2/k}$,

$$\int_0^r (1 + t^k)^{-2/k}\,dt \leq |f(z)| \leq \int_0^r (1 - t^k)^{-2/k}\,dt.$$

34. (I) $B_{nk+1} = \dfrac{1}{n!}\prod_{j=0}^{n-1}(j + 2/k) = \left|\binom{-2/k}{n}\right|$, (II) $B_{nk+1}/(nk+1)$.

MacGregor has obtained bounds if $f(z)$ is in ST or CV and $a_2 = a_3 = \cdots = a_k = 0$. He proves that the coefficients a_{nk+1} have the same bounds as in (I) and (II) and hence are sharp. He also obtains bounds for $|a_m|$ when $m \neq nk + 1$.
35. $(\tanh \pi/4)^{1/k}$.
36. One should compare this result with Gronwall 1916 [1-B-454], who

announced a different value for R_{CV}. Apparently Gronwall was in error.
38. Set $F(z) = z + \Sigma_2^\infty a_n z^n$ and $f(z) = z + \Sigma_2^\infty b_n z^n$. By a rotation assume that $\alpha = -\beta$, where $0 < \beta < \pi$. If $a_n \neq 0$ and $\sin n\beta = 0$, then there is *no* $f(z)$ that satisfies (66). Suppose $F(z) = z + z^2/2 \in ST$. If $|\cos \beta| < 1/2$, then $f(z)$ is not univalent. If $F(z) = k(z)$, then $f(z)$ is not univalent.
41. $f_1(z) = z + z^2/4 + z^3/48$. **42.** $z/(1 - cz)$, $|c| \leq 1$.
50. If $x = (5 - 2r + r^2)/8$, then $16\phi(r) = (r+1)^2(7 - 10r - r^2) = 0$, if $r = 4\sqrt{2} - 5$. Then $x = 9 - 6\sqrt{2} < 4\sqrt{2} - 5$.
51. Yes. $G(z) = z/(1 - z^2)$. **55.** Yes.
58. For $0 < Q_{ST}(z_0) \leq 1$ consider $f(z) = z + M$, with $M \geq 0$. For $Q_{ST}(z_0) > 1$, use $f(z) = z + Mz^n$ with n a suitable integer.
61. If $f(z)$ is in P', then $F(z) = z \exp \int_0^z \frac{f'(\zeta) - 1}{\zeta} d\zeta$ is in ST.

If $F(z)$ is in ST, then $f(z) = z + \ln \frac{F(z)}{z} - \int_0^z \ln \frac{F(\zeta)}{\zeta} d\zeta$ is in P'.

CHAPTER 9

2. If $g(z)$ is in $ST(1 - \alpha\beta)$ and $0 < \alpha, \beta < 1$, then $f(z) \equiv z(g(z)/z)^{1/\alpha}$ is in $ST(1 - \alpha)$.
6. $f(z) = \int_0^z e^{I(\zeta)} d\zeta$ where $I(z)$ is given by (97).
8. If $\Sigma_2^\infty n(n - \alpha)|a_n| \leq 1 - \alpha$ and $0 < \alpha < 1$, then $f(z)$ is in $CV(\alpha)$.
9. Yes. In Problem 7, take $0 < \epsilon < \alpha$, $f(z) = z - a_2 z^2$ where $a_2 = (1 - \alpha + \epsilon)/(2 - \alpha)$. In Problem 8 use equation (15).
18. (a) $\max \beta = 1 - \epsilon/2$, (b) $\max \beta = 1 - k\epsilon/2$.
19. Set $f(z) = z/(1 - z)^{2+\delta}$, with $\delta > 0$. Then $f'(z) = 0$ at $z = -1/(1 + \delta)$ in E.
23. $f(z) = ze^{I(z)}$ where $I(z) = \int_0^z \frac{(A - B)b(t)}{1 + Btb(t)} dt$ and $|b(z)| \leq 1$ in E.
25. $|b_{nk}| \leq 2(1 - \alpha)$. $P(z) = 1 + 2(1 - \alpha)z^k/(1 - z^k)$.
26. (a) The extremal function is $F(z) \equiv z/(1 - z^k)^{2(1-\alpha)/k}$.

(b) The extremal function is $\int_0^z [F(t)/t] dt$.

27. Yes, $f(z) = Cz$. **28.** $\ln z$ will enter the formula.
29. (a) $\frac{\ln(1 + cz)}{c}$, (b) $\frac{\sqrt{2}}{c}\left(\ln(1 + cz) - \frac{cz}{1 + cz}\right)^{1/2}$,

(c) $\dfrac{1}{4c}\ln^2(1+2cz+2\sqrt{c}\sqrt{z(1+cz)})$.

30. $\dfrac{z}{1+z}\left(\dfrac{3+z}{3(1+z)}\right)^{1/2}$. **34.** $\alpha = 0$.

36. One method is to use L'Hospital's Rule twice. **37.** Yes.

39. Set $\zeta = 0$. Yes. Set $f(z) = z/(1-z)$.

40. If $f(z)$ is in CV, then (A) $|1 - z/f(z)| \le |z|$, (B) $\operatorname{Re} f(z)/z \ge 1/2$, and (C) $r/(1+r) \le |f(z)| \le r/(1-r)$, with all three theorems sharp for $f(z) = z/(1-z)$.

47. $\alpha = \pi/4$. **48.** Try $z = 1 - e^{-2n\pi}$.

49. $|t| \le 1$, $\alpha = -\arg(t+1)$ unless $t = -1$.

50. We exclude the value $t = 0$. $f(z)$ is univalent if and only if t lies in either one of the two disks: $|t-1| \le 1$ and $|t+1| \le 1$. See Lemma 15.1.

51. Make a picture of the three disks.

54. $z/(1-z)$. Hint: First prove that $(f(z)-f(-z))/2$ is in $ST^{(2)}$ and then use Problem 8.34.

70. $f_1(z) = z^3$. If we require the normalization $f(z) = z + \cdots$, then set $f_2(z) = z - 2z^3$.

72. $|F(z_2) - F(z_1)| \ge \left|\int_{\mathscr{C}} c\, dz\right| - \left|\int_{\mathscr{C}} (F'(z) - c)\, dz\right| > 0$.

73. $\pi/2$.

74. Use $\zeta = 1/z$ and show that $|[F'(1/z) - 1]/z^2| \le 1$ in E. Then follow the method of Problem 72.

75. Use the Stieltjes integral formula from Problem 6, and first approximate $f(z)$ using a step function for $\mu(\phi)$. Equality if $f'(z) = 1/(1-z)^{2(1-\alpha)}$.

CHAPTER 10

2. Yes. **3.** $\tilde{k}_\phi(z) = zf'(z)$, $\mu_1 = \mu_2 = 1$, $\phi_2 = -\phi_1$.

4. See Problem 4.1.

5. Set $\lambda = \mu = 1/2$, $f(z) = z/(1 - ze^{i\phi})^2$, and $g(z) = z/(1 - ze^{-i\phi})^2$.

6. Consider $f(z) = z + 100z^2$. If $f(z)$ is in TR, then $f'(r) > 0$ in $(-1, 1)$.

9. Yes. **12.** Yes. It is also in $TR^{(n)}$. **13.** It is in TR.

14. Integrate along a radial line. **15.** It is in TR if $f(z)$ is.

16. If $f(z)$ is odd, then $f(z) = [f(z) - f(-z)]/2$.

17. $[k(z)-k(-z)]/2 = (z+z^3)/(1-z^2)^2$ is in TR.
19. See equation (26). **25.** No.
26. Use Theorem 10 and equation (35).
27. See Problem 8.10 and use Alexander's Theorem.
29. Set $t = (1-\cos\phi)/2$ and map E onto the cut plane.
30. If n is odd, $0 \le b_n \le 1$. If n is even, $-1 \le b_n \le 0$. Extreme functions are $F(z) \equiv z$ and $F(z) \equiv 1 - 1/z = \Sigma_1^\infty (-1)^{n+1}(z-1)^n$.
32. Let $\mu(t)$ have jumps of $1/2$ at $t = 0$ and $t = 1$. Then $2F(z) = z + 1/z$. Set $z_1 = ir$ and $z_2 = i/r$.
33. D is the common part of the two disks $|z-i| < \sqrt{2}$ and $|z+i| < \sqrt{2}$. One extremal function is $f(z) = (z+z^3)/(1-z^2)^2$.
34. $F(z) = tz/(1-z)^2 + (1-t)z/(1+z)^2$, $0 \le t \le 1$.
39. Use Problem 7.36 and Theorem 4. **40.** Set $F(z) = e^{-i\alpha}f(e^{i\alpha}z)$ in Problem 2.32.
42. Let $Q(\theta) = (1+r^2)\sin(\theta-\alpha) + r(3\sin\alpha - \sin(2\theta+\alpha))$. With $r = 0.55$ and $\alpha = 30°$ we find that $Q(0) = -0.10125 < 0$, $Q(180°) = 1.20125 > 0$, $Q(270°) \approx -0.028 < 0$, and $Q(300°) = 0.0725 > 0$.
48. $\nu = \pi$ and $\mu = \pi/2$ for both functions! **53.** See Problem 18.
54. No. Consider $f(z) = z + z^3$ in $TR^{(2)}$.
55. Yes, $f(z) = z/(1 \pm z^2)$. **56.** See Problem 24.
57. Use equation (24). Yes. Set
$$g(z) = \frac{1}{2}\left(\frac{z}{(1+z)^2} + \frac{z}{(1-z)^2}\right) = z\,\frac{1+z^2}{(1-z^2)^2} \text{ in } TR^{(2)}$$
and $f(z) = \sqrt{g(z^2)} = z\sqrt{(1+z^4)/(1-z^4)} = z + (3/2)z^5 + \cdots$.
59. Set $f(z) = [(z+a)^2 - a^2]/2a$ with $a = 1/4$.
67. Set $f(z) = z/(1+z)^2$ and let $z \to i$.
68. Set $f(z) = z/(1+z)$ and let $z \to i$.
69. $|a_n| \le n/[1+(n-1)\alpha]$, and this is sharp for each n.
70. First prove that $(1-z^2)f'(z) = (f'(0) + \overline{f'(0)}b(z))/(1-b(z))$ for some $b(z)$ in B_0.
71. Set

$$H(z) \equiv G(e^{-i\alpha}z) = \frac{e^{-i\alpha}z - e^{i\alpha}z^2}{(1-z)^3}.$$

Then $H(e^{i\theta}) = -\sin(\alpha+\theta/2)/4\sin^3(\theta/2)$ is real. Further $H'(z) = 0$ for $\zeta = e^{-i\alpha}(\pm\sqrt{1-4\sin^2\alpha} - 2i\sin\alpha)$ and $|\zeta| = 1$ if $0 < \alpha < \pi/6$.

INDEX OF SPECIAL SYMBOLS

Some standard symbols are not included in this list. In some cases we omit a symbol if it is used locally in a very small part of the book. No page number is given for certain standard symbols.

SETS OF NUMBERS OR POINTS

Symbol	Description or Meaning	Page
C_r	$\|z\| = r$	27
D	A domain, open connected set	5
∂D	The boundary of D	
$f(D)$	The image of D under $f(z)$	12
Γ	A curve	7
Γ_r	$f(C_r)$	29
$[a, b]$	$a \leq x \leq b$	
(a, b)	$a < x < b$	
L	A line $w = A + e^{i\alpha}t$	77
H^+, H^-	$\text{Re } w > 0$, $\text{Re } w < 0$	77
Γ_z, Γ_w	A curve in the z-plane (w-plane)	108
γ_t	An arc of a circle	200
C	A convex set	93
$E(C)$	The extreme points of C	95
$D_2(i)$	A special type of domain	199
$D_1^+(i)$	A special type of domain	202
$D_1^-(i)$	A special type of domain	202
$\mathscr{R}(\mathscr{M}(z_0))$	The set of values at z_0 for all $f(z)$ in $\mathscr{M}$	188

SETS OF FUNCTIONS

If XY denotes a set of functions, the set XYR denotes the subset of functions in XY for which the coefficients in the series expansion at the origin (or at ∞) are all real. For example, SR, STR, CVR, PR.

The symbol $XY^{(k)}$ denotes the subset of functions in XY that have k-fold symmetry about the origin (or about ∞). For example, $S^{(k)}$, $SR^{(k)}$, $CV^{(k)}$, $CVR^{(k)}$. For $P^{(k)}$ see Problem 7.47.

The symbol XY' denotes a set of normalized functions $f(z)$ such that the derivative $f'(z)$ is in XY.

The symbol $\widetilde{XY}$ denotes a set that has some characteristic in common with XY but is not identical with XY.

The symbol $XY(\alpha)$ is the subset of functions in XY that are of order α. For example, $ST(\alpha)$, $CV(\alpha)$. But $SP(\alpha)$ denotes a spirallike function of type α.

If XY contains functions that are not univalent in E, then SXY denotes the subset of functions in XY that are univalent in E.

XY^x denotes the set of functions defined in E^x: $1 < |\zeta| < \infty$, which has the same character as the set XY defined in E .

This system can become very unwieldy. For example we may have a set denoted by $SCSTR^{(k)x}$, which by our agreement is the set of k-fold symmetric functions that are univalent, close-to-star, and real on the part of the real axis in $1 < |\zeta| < \infty$.

Each set has the normalization that is specified in the definition given in the reference.

SETS OF FUNCTIONS

SPECIAL FUNCTIONS AND FUNCTIONALS

Symbol	Description or Meaning	Page
$K(w)$	The inverse of $k(z)$	37
$\widetilde{k}_\phi(z)$	$z/(1-bz+z^2)$, $b = 2\cos\phi$	35
$k_p(z)$	$z/(1-bz+z^2)$, $b = p+1/p > 2$	22
$k^\star(z, \alpha)$	$z/(1-z)^{2(1-\alpha)}$	140
$k(z, M)$	$M\dfrac{M(1-z)^2+2z-(1-z)\sqrt{M^2(1-z)^2+4Mz}}{2z}$	38
$k(c, z)$	$\dfrac{1}{2c}\left[\left(\dfrac{1+z}{1-z}\right)^c - 1\right]$	22
$k(z; \alpha)$	$\left[\dfrac{1}{\alpha}\displaystyle\int_0^z \dfrac{t^{1/\alpha}\,dt}{t(1-t)^{2/\alpha}}\right]^\alpha$, $\alpha > 0$	146
$K(z, \phi)$	A kernel function	189
$A(r)$, $A(r, f)$	Area of $f(E_r)$	25
$L(r, f)$	Length of $f(C_r)$	32
$L(\alpha, r, f)$	Length of $f(te^{i\alpha})$, $0 \le t \le r$	32
$L_0(z)$	$(1+z)/(1-z)$	78
$M(r, f)$	$\max \|f(re^{i\theta})\|$, r fixed	31
$m(r, f)$	$\min \|f(re^{i\theta})\|$, r fixed	31
$Q_{ST}(z)$	$zf'(z)/f(z)$	137
$Q_{ST}(f)$	Same as $Q_{ST}(z)$	137
$Q_{CV}(z)$	$1 + zf''(z)/f'(z)$	137
$Q_{CV}(f)$	Same as $Q_{CV}(z)$	137
$J[f]$	An arbitrary functional	42
$\phi(z)$	$e^{i\alpha}\dfrac{z-a}{1-\bar{a}z}$	10
$f[z_1, \ldots, z_n]$	nth order divided difference	90
$D^\star(k, n)$	$\displaystyle\prod_{\substack{j=1\\ j\neq k}}^{n} (z_j - z_k)$	90
$\Delta(re^{i\theta})$	$\dfrac{f(z)-f(\bar{z})}{z-\bar{z}}$	182
$b(z)$	A function in B_0	84
P_n	nth Legendre polynomial	213
$\mu(\phi)$	Nondecreasing function	79
$S_n^{(1)}(z)$	$\dfrac{1}{n}[nz + (n-1)z^2 + \cdots + 2z^{n-1} + z^n$	156

Symbol	Description or Meaning	Page
$S_n^{(2)}(z)$	$\dfrac{2}{n(n+1)} \sum_{k=1}^{n} \binom{n+2-k}{2} z^k$	156
$S_n^{(3)}(z)$	$\dfrac{6}{n(n+1)(n+2)} \sum_{k=1}^{n} \binom{n+3-k}{3} z^k$	157
$f \circ g$	Composite function, $f(g(z))$	Never used

SPECIAL NUMBERS OR QUANTITIES

Symbol	Description or Meaning	Page
γ	A value omitted by a function	19
R_S	Radius of univalence	22
$R_{\mathscr{P}}$	Radius for the property $\mathscr{P}$	119
τ	Argument of a tangent line	109
$K(\Gamma)$	Curvature	133
η	$\lvert\eta\rvert = e^{2\pi i/N}$	3
x	$e^{-i\theta}$	159

MISCELLANEOUS SYMBOLS

Symbol	Description or Meaning	Page
PLE	Proof is left as an exercise	x
PWO	Proof will be omitted	x
$f(z) << F(z)$	$f(z)$ is dominated by $F(z)$	82
$f(z) \prec F(z)$	$f(z)$ is subordinate to $F(z)$	85
★	A more difficult problem	

INDEX OF NAMES

INDEX OF TOPICS